Markov Chains

Markov Chains

From Theory to Implementation and Experimentation

Paul A. Gagniuc
University Politehnica of Bucharest, Faculty of Engineering in Foreign Languages

Registered Offices
John Wiley & Sons, Inc., 111 River Street, Hoboken, NJ 07030, USA

Editorial Office
111 River Street, Hoboken, NJ 07030, USA

For details of our global editorial offices, customer services, and more information about Wiley products visit us at www.wiley.com.

Wiley also publishes its books in a variety of electronic formats and by print-on-demand. Some content that appears in standard print versions of this book may not be available in other formats.

Library of Congress Cataloging-in-Publication Data

Names: Gagniuc, Paul A., author.
Title: Markov chains : from theory to implementation and experimentation / Paul A. Gagniuc.
Description: Hoboken, NJ : John Wiley & Sons, 2017. | Includes bibliographical references and index. |
Identifiers: LCCN 2017011637 (print) | LCCN 2017018061 (ebook) | ISBN 9781119387572 (pdf) | ISBN 9781119387589 (epub) | ISBN 9781119387558 (cloth)
Subjects: LCSH: Markov processes.
Classification: LCC QA274.7 (ebook) | LCC QA274.7 .G34 2017 (print) | DDC 519.2/33–dc23
LC record available at https://lccn.loc.gov/2017011637

Cover image: © KTSDESIGN/SCIENCE PHOTO LIBRARY/Gettyimages
Cover design by Wiley

Set in 10/12pt WarnockPro by Aptara Inc., New Delhi, India

10 9 8 7 6 5 4 3 2 1

I dedicate this book to my children
– Nichita-Constantin and Ana-Sofia

Contents

Abstract *ix*
Preface *xi*
Acknowledgments *xiii*
About the Companion Website *xv*

1 Historical Notes *1*
1.1 Introduction *1*
1.2 On the Wings of Dependent Variables *2*
1.3 From Bernoulli to Markov *5*

2 From Observation to Simulation *9*
2.1 Introduction *9*
2.2 Stochastic Matrices *9*
2.3 Transition Probabilities *11*
2.4 The Simulation of a Two-State Markov Chain *14*

3 Building the Stochastic Matrix *25*
3.1 Introduction *25*
3.2 Building a Stochastic Matrix from Events *25*
3.3 Building a Stochastic Matrix from Percentages *32*

4 Predictions Using Two-State Markov Chains *37*
4.1 Introduction *37*
4.2 Performing the Predictions by Using the Stochastic Matrix *37*
4.3 The Steady State of a Markov Chain *46*
4.4 The Long-Run Distribution of a Markov Chain *55*

5 Predictions Using *n*-State Markov Chains *61*
5.1 Introduction *61*
5.2 Predictions by Using the Three-State Markov Chain *61*

5.3 Predictions by Using the Four-State Markov Chain *71*
5.4 Predictions by Using *n*-State Markov Chains *80*
5.5 Markov Chain Modeling on Measurements *84*

6 Absorbing Markov Chains *93*
6.1 Introduction *93*
6.2 The Absorbing State *93*

7 The Average Time Spent in Each State *99*
7.1 Introduction *99*
7.2 The Proportion of Balls in the System *99*
7.3 The Average Time Spent in A Particular State *100*
7.4 Exemplification of the Average Time and Proportions *101*

8 Discussions on Different Configurations of Chains *107*
8.1 Introduction *107*
8.2 Examples of Two-State Diagrams *113*
8.3 Examples of Three-State Diagrams *115*
8.4 Examples of Four-State Diagrams *117*
8.5 Examples of State Diagrams Divided into Classes *123*
8.6 Examples of State Diagrams with Absorbing States *127*
8.7 The Gambler's Ruin *128*

9 The Simulation of an *n*-State Markov Chain *131*
9.1 Introduction *131*
9.2 The Simulation of Behavior *131*
9.3 Simulation of Different Chain Configurations *145*

A Supporting Algorithms in PHP *165*

B Supporting Algorithms in Javascript *193*

C Syntax Equivalence between Languages *223*

Glossary *225*
References *227*
Index *231*

Abstract

Independent trials are the basis of statistics and classical probability theory. Nevertheless, in modern probability theory, the chance processes are viewed through the prism of past events, in which previous outcomes have an influence on predictions of future events. In this regard, Markov chains are central to the understanding of random processes and can be used to model systems at various scales. Markov chains have found many applications. Some of the most known applications are those from meteorology (day-to-day rainfall prediction), biology (population dynamics), chemistry (chemical reactions), bioinformatics (DNA analysis or protein folding), and information technology (i.e. speech recognition). Although the Markov chain method is used in many areas of science and engineering, it is yet to be used in novel approaches that are compatible with this type of analysis. The main theme of the book revolves around examples based on objects such as jars (representing states) and balls (representing transition probabilities) of various colors. Thus, any type of Markov chain configuration is explained in terms of real experiments. These examples relate to each other from chapter to chapter enabling a gradual understanding of the phenomena. The theory is also accompanied by an algorithm implementation for each example. Chapter 1 begins with a general introduction into the history of probability theory, covering different time periods. In this chapter, the introduction to discrete-time is made using quantifiable examples showing how the field of probability theory arrived in recent times at the notion of dependent variables (Markov model) from experiments related to independent variables (Bernoulli model). Chapters 2 and 3 are an introduction to simple stochastic matrices and transition probabilities followed by a simulation of a two-state Markov chain. The description starts from the observation of events within a system up to the simulation of that system through a Markov chain. The construction of a stochastic matrix is shown based on both a sequence of observations and observations provided in percentages. Chapter 4 begins with an introduction to predictions that use a two-state Markov chain. Here, the notion of steady state is first approached and discussed in connection with the long-run distribution behavior of the Markov chain. Chapter 5 describes some examples by considering predictions based on Markov chains with more than two states. The first two examples include a three-state Markov chain and a four-state Markov chain, after which a gradual generalization is made for an arbitrary number of states (n-states). In Chapter 6, the notion of absorbing Markov chains it is approached by using tangible examples. Chapter 7 covers a topic linked to the average time spent in a state, whereas Chapter 8 covers discussions on different configurations of chains. Different configurations of state diagrams provide solutions for different problems in many fields. As a continuation, Chapter 9 covers the simulation of an n-state Markov chain used for verifying experiments of various diagram configurations. Overall, the book intends a completely different approach on Markov chains, which is based on four convergent lines that include mathematics, implementation, simulation, and experimentation.

Preface

The concepts of probability and stochastic theory are being successfully used in many fields. Some of the most known applications of Markov chains are those from meteorology (day-to-day rainfall prediction), biology (population dynamics), gambling (studies related to chance), chemistry (chemical reactions), bioinformatics (DNA analysis or protein folding), and information technology (speech recognition). It is also the preferred method of prediction in economics and social sciences, and forms one of the bases of quantum mechanics. However, Markov chains are somehow restricted to elitist groups, which unintentionally slow an extensive use of this mathematical model for more worldly developments. The book addresses Ph.D. students and college students from various fields that wish to quickly understand the step-by-step logic behind the Markov model. Moreover, historical notes shed light on the roots of the important ideas that led to the Markov model development and its subsequent variants. One of the themes of the book takes into account different configurations of Markov chains and their limitations. Many of these examples are presented from simple to complex in a comparative manner with an express use of color graphics. Thus, the book does not require prior extensive knowledge in probability theory. Nevertheless, the understanding of Markov chain process involves a certain level of abstract imagination from the one who intends to use the method for particular purposes. The primary objective of this book is to initiate both researchers and students in the art of stochastic modeling in an attractive manner. There are perhaps many important subtle techniques, which are not covered here. This book focuses on both theory (mathematical aspects) and implementation (programming issues) in a common conceptual framework. The prediction based on previous events is the main theme of the book. The book allows the analysis of the algorithms in three different programming or script languages, namely, Visual Basic, Java Script, and PHP. The implementation of the algorithms represents both an exercise for the reader and a method of step-by-step verification of the data shown in the supporting theory of the book.

Acknowledgments

I would like to express my gratitude to Ani Apati for the original representation of the portraits of Gerolamo Cardano, Jakob Bernoulli, and Andrei Markov, made specifically for this book.

About the Companion Website

This book is accompanied by a companion website:

www.wiley.com/go/gagniuc/markovchains

The website includes:

Complete open source applications
- Markov Chains – Prediction framework
- Markov Chains – Simulation framework
- Markov Chains – The weather

Fifty-one ready to use algorithms
- 17 ready to use algorithms in JavaScript as HTML files
- 17 ready to use algorithms in PHP as script files
- 17 ready to use algorithms in Visual Basic as BAS files

1

Historical Notes

1.1 Introduction

Probability is the measure of how likely a future event is. It is not by "chance" that most of the examples related to the understanding of probability are connected to objects like dices, cards, or coins. Historical events show that most primitive attempts on probability theory have the roots in gambling [1]. Given the implications that gambling had over time, particularly the social consequences of it, great efforts have been made to avoid or understand uncertainty. Historians have looked to Aristotle and beyond when searching for the origins of the probabilistic concepts [1]. The very first ideas for this fundamental principle may derive directly from Aristotle's Ethics, where the concept of "justice" took new forms over time [1]. Later, the medieval poem *De Vetula* ("On the Old Woman") appeared around the year 1250. This poem is a long thirteenth-century elegiac comedy that contains first written references on gambling [2]. The non-poetic content of *De Vetula* makes references to the connection between the number of combinations and the expected frequency of a given total [2]. Gerolamo Cardano (1501–1575) has made the first written references in defining odds as the ratio of favorable to unfavorable outcomes [1]. In his *Liber de Ludo Aleae* ("Book on Games of Chance") published in 1564 or later, Cardano was among the first to approach probabilities in games of chance [1]. A few decades later, uncertain events related to gambling resulted in the well-known mathematical theory of probability formulated by Pierre de Fermat and Blaise Pascal (1654) [3]. Just 3 years later in 1657, Christian Huygens (1629–1695) published a dedicated book on probability theory related to problems associated with games of chance, entitled *De Ratiociniis in Ludo Aleae* ("On Reasoning in Games of Chance") [4, 5]. A milestone contribution of Jakob Bernoulli (1654–1705) in probability theory was published post-mortem in 1713, under the title *Ars conjectandi* ("The Art of Conjecturing") [6, 7]. Bernoulli was concerned with predicting the probability of unknown events [7]. In his work Bernoulli describes what today is known as the weak law of large numbers [7]. This law shows that the average of the results obtained from an

Markov Chains: From Theory to Implementation and Experimentation, First Edition. Paul A. Gagniuc.
© 2017 John Wiley & Sons, Inc. Published 2017 by John Wiley & Sons, Inc.
Companion website: www.wiley.com/go/gagniuc/markovchains

Gerolamo Cardano (1501–1575) Jakob Bernoulli (1654–1705) Andrei Markov (1856–1922)

increasing number of trials should converge toward the expected value [7]. For instance, consider the flipping of an unbiased coin. As the number of flips goes to infinity, the proportion of heads will approach 1/2. Let us consider another example: without our knowledge, x black balls and y white balls are placed in a jar (Figure 1.1a). To determine the proportions of black balls and white balls from the jar by experiment, a series of random draws must be performed. Whenever a black ball or a white ball is drawn, the observation is noted. The expected value of white versus black observations will converge toward the real ratio from the jar as the number of extractions increases. Therefore, Bernoulli proved that after many random draws (trials), the observations of white balls versus black balls will converge toward the real ratio of black balls versus white balls from the jar. Almost 100 years after Bernoulli, Pierre de Laplace (1749–1827) severs the thinking monopoly that gambling had on the probability theory [1–7]. In 1812, Laplace publishes the *Théorie Analytique des Probabilités* ("Analytical Theory of Probability") in which it introduces probability to general sciences [8, 9].

1.2 On the Wings of Dependent Variables

Another milestone in probability theory was made almost 100 years later by Andrei Markov (1856–1922) [10, 11]. In the Bernoulli model, the outcome of previous events does not change the outcome of current or future events. Today it is obvious that the events are not independent in many cases; however, in the past, this was not that obvious (in the mathematical sense). The term of *dependent events*, or *dependent variables*, refers to those situations when the probability of an event is conditioned by the events that took place in the past. A colleague of Markov, namely Pavel Nekrasov, assumed that independence is a condition for the law of large numbers [12]. Following a dispute with Pavel Nekrasov, Markov considered that the law of large numbers

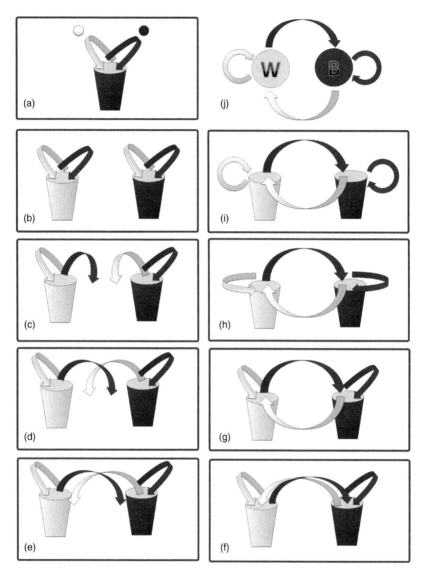

Figure 1.1 From Bernoulli model to the Markov model. (a) The Bernoulli model with a single jar filled with different proportions of black and white balls. The curved arrows show from which jar the extraction was done and to which jar the next draw will be made. Black curved arrows show the extraction and observation of black balls whereas white curved arrows show the extraction and observation of white balls. (b) Two Bernoulli jars. A white jar and a black jar, each filled with different proportions of black and white balls. Here, draws are still independent from one another. (c–f) Shows how dependence occurs between the two jars if the color of the curved arrows is "attracted" to the color of the jars. (g–j) Shows how the two jars system is transformed into a Markov diagram by changing the angle of viewing of the jars, from the side view to the top view.

can be also valid in the case of dependent variables. In 1906, Markov extends Bernoulli's results to dependent variables and began developing his reasoning about chains of linked probabilities (Figures 1.1a–j) [10]. Markov's connection with Bernoulli it is indirectly but deeply rooted in the history of the Academy of Sciences in St. Petersburg. Prior to Markov's time, the Academy included none other than the great Leonhard Euler (1707–1783) and the sons of Jakob Bernoulli, namely Nicholas Bernoulli (1687–1759) and Daniel Bernoulli (1700–1782) [12]. In 1907, Markov proved that the independence of random variables was not a required condition for the validity of the weak law of large numbers and the central limit theorem [10, 11]. For his demonstration, he envisioned a virtual machine (Figures 1.1f and 1.1j).

Let us consider two jars which represent the two states of a machine (Figure 1.1b). One is painted in black (state 1) and the other is painted in white (state 0). Both contain certain proportions of both white and black balls. First, an extraction of a ball from one of the jars is made; let us choose the black jar (state 1). If the black ball is drawn, then the next draw is made again from the black jar (state 1). If the white ball is drawn, then the next draw is made from the white jar (state 0). Let us consider that a white ball has been pulled. Therefore, the next draw is made from the white jar. If a white ball is drawn from the white jar, then the next draw is made again from the white jar (state 1). If a black ball is drawn from the white jar then the next draw is made from the black jar (state 0). Thus, these events may continue indefinitely. What can be immediately noticed is that the current extraction is dependent on the previous extraction. As long as both states of the machine are reachable (each jar contains both white and black balls), the number of visits to each jar will converge, as in the Bernoulli model, to a specific ratio. By this simple example, Markov showed that the law of large numbers applies in the case of dependent variables. But what does "both states are reachable" mean? If the black jar has only black balls inside, then all drawings are made from the black jar; therefore, the white jar is unreachable. However, what if draws are first started from the white jar? Eventually, after a number of draws, a black ball is drawn from the white jar. Once the black ball is drawn from the white jar, the next draw will be made from the black jar. Since the black jar contains in this case only black balls, from this point forward, all future draws will be made only from the black jar. Therefore, the white jar will be unreachable. In order to make the white jar reachable, the black jar must contain at least one white ball from the total number of balls (n). That one white ball it will provide a very small chance ($1/n > 0$), of switching the extraction of balls from the black jar to the white jar. Taking these observations into account, the probability of extracting a white ball from the black jar will be:

$$P[\text{white}] = \frac{1}{n}$$

Whereas the probability of extracting a black ball from the black jar will be:

$$P[\text{black}] = \frac{(n-1)}{n}$$

Also notice that:

$$P[\text{white}] + P[\text{black}] = \left(\frac{1}{n}\right) + \frac{(n-1)}{n} = 1$$

In 1913, by pencil and paper, Markov applied his method for a linguistic analysis of the first 20,000 letters from one of Pushkin's poems [12]. Thus, he showed that the letter probabilities in Pushkin's poem are not independent. This linguistic analysis sparked the interest of many scientists at that time and quickly brought a worldwide revolution in science and technology [12]. Many great minds preoccupied by uncertainty made their contribution over time to the probability theory. Nevertheless, what had begun as an analysis of gambling rooted in decadence is now the main weapon used for the progress of mankind.

1.3 From Bernoulli to Markov

Simple exemplifications are crucial for understanding the Markov process. A stochastic process is visually represented by state diagrams. Circles inside a diagram represent states while arrows indicate the probability of moving from one state to another (Figure 1.1j). In our days, state diagrams are usually taken for granted without any possible roots underlying these types of visual representations. As stated above, A. Markov was influenced by Jacob Bernoulli [10]. Therefore, let us start from Bernoulli's jar (Figure 1.1a). Consider an opaque jar filled with white and black balls whose proportions are unknown. Of course, a simple counting of the balls is not allowed. What is allowed is a repeated extraction of balls from the jar and the notation on paper of how many balls are white and how many balls are black. The question is: can a similar jar be filled with the same proportion of white and black balls? The expected value of white versus black observations will converge toward the real ratio from the jar, as the number of extractions increases. Also notice that in Bernoulli model, the outcome of previous events does not change the outcome of current or future events. Thus, all draws are independent of each other. Let us consider that in this Bernoulli model, the aim is to find out what is the probability of a specific sequence of draws, for instance black ball, white ball, black ball, black ball: $P[\text{BWBB}]$. If these four independent events (BWBB) have probabilities $P[\text{B}]$ and $P[\text{W}]$, the joint probability of all four events is the product:

$$P[\text{BWBB}] = P[\text{B}] \times P[\text{W}] \times P[\text{B}] \times P[\text{B}]$$

In other words, the probabilities are not linked. But what does that mean? How can probabilities be linked? Consider two jars which represent the two states of a machine (Figure 1.1b). One jar is painted in black (state B) and the other jar is painted in white (state W). Both jars contain white balls and black balls in unknown proportions. Thus, the percentage of white versus black balls in each jar is the aim of this example. What can be done? Consider that two individuals are involved in an experiment. These two individuals are named Alice and Bob. Presumably, a first draw is made from one of the jars by Alice. A draw rule is also imposed, namely the color of the current ball indicates the color of the jar from which the next draw will be made (Figure 1.1f). By following this rule, suppose 80 draws are made by Alice behind a screen. Therefore, the interplay between jars (states) is not observable by Bob. However, Alice shows Bob the ball each time a black ball is drawn from the black jar, or a white ball is drawn from the white jar. The color of the ball is written on paper by Bob each time Alice shows him a ball. Thus, at the end of the 80th draw, Bob notes that Alice showed him the black ball 20 times and the white ball 30 times. Thus, Alice asks Bob to tell her the proportion of balls in each jar based exclusively on these observations. Taking into account the draw rule, Bob immediately realizes that each time Alice showed him the black ball she returned to the black jar, and each time she showed him the white ball she returned to the white jar. Initially, Bob imagines the jars as independent systems (Figures 1.1b and 1.1c). When regarded as an independent Bernoulli experiment with a single jar, the proportion of balls can be easily determined. Thus, the probability to draw a black ball ($P_{\text{white jar}}[B]$) or a white ball ($P_{\text{white jar}}[W]$) from the white jar equals 1:

$$P_{\text{white jar}}[B] + P_{\text{white jar}}[W] = 1$$

If the probability of extracting a white ball ($P_{\text{white jar}}[W]$) from the white jar is known, then the probability of extracting a black ball from the white jar can also be found, namely (Figure 1.2b):

$$P_{\text{white jar}}[B] = 1 - P_{\text{white jar}}[W]$$

Similarly, the same rules apply to the black jar. The probability of extracting a black or a white ball from the black jar is also equal to 1 (Figure 1.2c). Taking into consideration the above properties, Bob wonders how he can approximate the proportion of white versus black balls in each jar. In order to find out the proportion of balls from the two jars, he must deduce the unknown probabilities from the two jars, namely $P_{\text{black jar}}[W]$ and $P_{\text{white jar}}[B]$ (Table 1.1). First, it starts from what he knows and divides the number of observations by the total number of draws to obtain $P_{\text{black jar}}[B]$ and $P_{\text{white jar}}[W]$ (Table 1.1). Secondly, it uses the expression $1 - P_{\text{white jar}}[W]$ to find $P_{\text{white jar}}[B]$ and the expression $1 - P_{\text{black jar}}[B]$ to find $P_{\text{black jar}}[W]$ (Table 1.1). In the final step, Bob transforms the probabilities into percentages by multiplying the probability values with 100. Bob then shows his findings to Alice in the form of proportions: the white jar contains ~62.5% black balls and 37.5% white balls whereas the black jar contains ~75% white balls and 25% black balls. The value of $P_{\text{black jar}}[B]$ represents

(a)

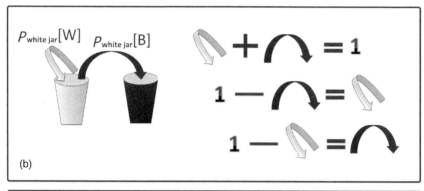

(b)

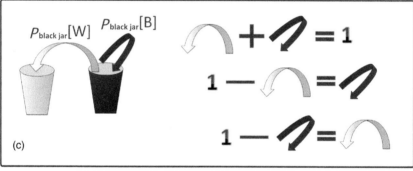

(c)

Figure 1.2 The two jars seen independently when disconnected from the Markov diagram. (a) The Markov model represented by two jars. (b) The white jar observed independently from the black jar. (c) The black jar observed independently from the white jar. The black arrows signify the transitions to the black jar, and the white arrows signify the transitions to the white jar.

the black balls in the black jar and the value of $P_{\text{white jar}}[W]$ represents the white balls in the white jar. Therefore, in this case, $P_{\text{white jar}}[B]$ and $P_{\text{black jar}}[W]$ counterparts represent the interplay between jars (states). Consequently, what was initially hidden behind the screen by Alice is now clear to Bob. Approximately 62.5% of the time the extractions were shifted from the white to the black jar and ~75% of the time the extractions were shifted from the black to the white jar. In other words, the number of trips from the white jar to the black jar represents the proportion of black balls inside the white jar (Figure 1.2b). In contrast,

Table 1.1 Finding the unknown probabilities of the two jars.

In the black jar	In the white jar
$P_{\text{black jar}}[B] = 20/80 = 0.25$	$P_{\text{white jar}}[W] = 30/80 = 0.375$
$P_{\text{black jar}}[W] = 1 - P_{\text{black jar}}[B]$	$P_{\text{white jar}}[B] = 1 - P_{\text{white jar}}[W]$
$P_{\text{black jar}}[W] = 1 - 0.25$	$P_{\text{white jar}}[B] = 1 - 0.375$
$P_{\text{black jar}}[W] = 0.75$	$P_{\text{white jar}}[B] = 0.625$

the number of trips from the black jar to the white jar represents the proportion of white balls inside the black jar (Figure 1.2c). When the Markov rule applies, the two jars have linked probabilities. Thus, at least one parameter of each jar (state) must be known in order to rebuild the unknown transition probabilities from the system (Figures 1.2b and 1.2c).

2

From Observation to Simulation

2.1 Introduction

A two-state Markov chain is the most basic model which can be used for the illustration of the Markov process. Based on this simple model, a more complex framework can be developed for dedicated purposes. Therefore, in the first instance, this chapter makes a more detailed introduction to stochastic matrices and transition probabilities discussed briefly in Chapter 1. In a second step, the theory is followed by a simulation of a two-state Markov chain.

2.2 Stochastic Matrices

Stochastic matrices have found use in probability theory, computer science, economics, chemistry, biology as well as bioinformatics and population genetics [13–16]. In fact, stochastic matrices can be used for predictions in many scientific areas, but also for problems that revolve around everyday life. A stochastic matrix or a Markov matrix (Figure 2.1j), is a matrix used to describe the transitions of a Markov chain (Figure 2.1a). The stochastic matrix is known by many names such as a probability matrix, a transition matrix, or a Markov matrix. Each value stored in this type of matrix is a nonnegative real number representing a probability [17]. Therefore, a stochastic matrix is composed of stochastic vectors (also named probability vectors) which store the possible outcomes of an event (a discrete random variable). In order to build a stochastic matrix, the stochastic vectors are positioned one above the other or side by side (Figure 2.2c). Thus, three types of stochastic matrices can occur. (1) The right stochastic matrix. A right stochastic matrix is a square matrix of nonnegative real numbers, with each row summing to 1. Each row of a right stochastic matrix is a stochastic vector (Figure 2.2a). (2) The left stochastic matrix is a square matrix of nonnegative real numbers, with each column summing to 1. Thus, each column of a left stochastic matrix is a stochastic vector (Figure 2.2b). (3) The doubly stochastic matrix. The doubly stochastic matrix is

Markov Chains: From Theory to Implementation and Experimentation, First Edition. Paul A. Gagniuc.
© 2017 John Wiley & Sons, Inc. Published 2017 by John Wiley & Sons, Inc.
Companion website: www.wiley.com/go/gagniuc/markovchains

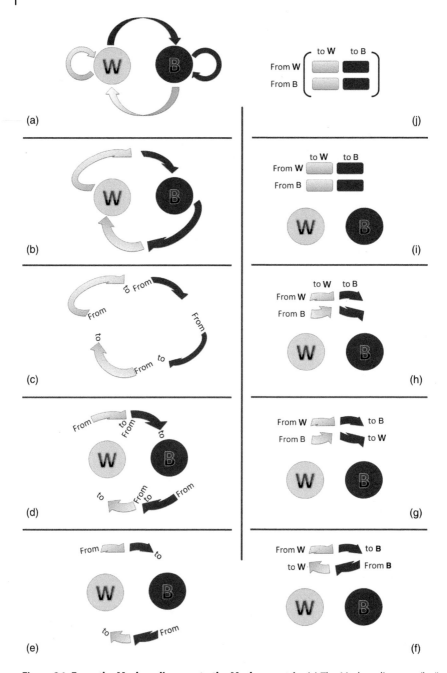

Figure 2.1 From the Markov diagram to the Markov matrix. (a) The Markov diagram. (b–i) Gradual deformation of the Markov diagram arrows until they are taking the form of the matrix cells. (j) The transition matrix, also named the Markov matrix.

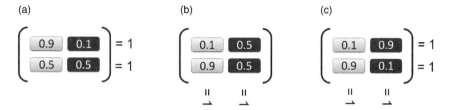

Figure 2.2 Types of stochastic matrices. (a) A right stochastic matrix, in which values from rows sum up to 1; (b) a left stochastic matrix, in which values from columns sum up to 1. (c) A doubly stochastic matrix, in which values from both rows and columns sum up to 1.

a square matrix of nonnegative real numbers with each row and column summing to 1 (Figure 2.2c). Note that a right stochastic matrix with row probability vectors is used throughout the examples given here (Figures 2.2a and 2.1).

2.3 Transition Probabilities

Consider two jars which represent the two states of a machine (Figure 1.2a or Figure 2.1a). One is painted in black (state B) and one is painted in white (state W). Both contain certain proportions of both white and black balls (Figure 2.1a). By respecting the chain rule, a series of draws can be made. In the chain rule, the color of the current ball indicates the color of the jar from which the next draw is made. By using a single letter, the color of the ball from each draw can be written on paper. For instance, letter "W" can be written when a white ball is drawn or letter "B" can be written when a black ball is drawn. At the end of an experiment, draws are stopped and a long string composed of "W" and "B" letters is obtained. For example, a number of 16 draws generate a sequence of 16 letters. Such a hypothetical sequence may look like: "WWBBWBWBBWBWBBBWW". Now the question arises: with the string of observations supplied above, is it possible to quantify how often draws jump from one jar to the other jar? Or how often a certain type of ball is drawn from a specific jar? The transitions between letters can be counted upon a careful visual inspection of the sequence. These transitions can take a limited number of forms, such as: WB, WW, BB, and BW. In order to see how often draws jump from one jar to the other jar, the *transition probability* between letters must be found. Note that $P[A|B]$ notation signifies *a transition from "A" to "B"* (see Glossary). Two key elements are of major importance when a transition probability is calculated from a sequence of events. (1) The pair of letters (or events) that denotes a transition (i.e., the total number of "WB" pair of letters in the sequence), and (2) the total number of the first letter in that pair (i.e., the total number of "W" letters in the sequence). First, the transitions between letter "W" and letter "B" are colored in orange to highlight the pair of letters in the sequence, as follows: "W~~W~~BWBWBBWBWBBBWW". A simple counting shows five transitions from letter "W" to letter "B". This observation implies that a black ball has been drawn five times from the white jar. This further indicates that a transition from the white jar to the black jar was made five times. The first letter of the pair is also counted. Thus, letter "W" is colored in orange to highlight the single letters, as follows: "W~~W~~BWBWBBWBWBBBW~~W~~". The count shows a total of six letters "W". The rule dictates that regardless of the jar from which the draw was made, if the current ball is white, the next draw is made from the white jar. Therefore, this last observation indicates that the transition toward the white jar has been made six times. Next, the transition probability from letter "W" to letter "B" (noted as $P_W[W|B]$) is calculated. Thus, the number of pairs is divided by the number of single letters:

$$P_W[W|B] = \frac{5}{6} = 0.833$$

Note: The *first* and *last* letters from the sequence are not taken into account when calculating the transition probabilities ("W~~W~~BWBWBBWBWBBBW~~W~~"). The first jar (state) is chosen by the experimenter and is not a part of the stochastic process. After the last draw, the transition is made to nowhere. Therefore, the last state to which the transition should be made is unknown. Also, the first transition (pair of letters) in the sequence cannot be counted since it is made from the first letter (state) that has been chosen by the experimenter ("W~~W~~BWBWBBWBWBBBWW"). To find out the value of $P_W[W|W]$, the determination of $P_W[W|B]$ should be sufficient (Figure 1.2). The probability of a transition from letter "W" to letter "B" is 0.833. Therefore, the probability of a transition between letter "W" to letter "W" should be:

$$P_W[W|W] = 1 - P_W[W|B] = 0.166$$

However, this result (the $P_W[W|W]$ value) can be verified by analyzing the sequence further in the same manner. Inside the sequence, one transition (pair of letters) is counted from letter "W" to letter "W" ("W~~W~~BWBWBBWBWBBBWW"). Also, the first letter "W" of the pair ("WW") appears six times ("W~~W~~BWBWBBWBWBBBW~~W~~"):

$$P_W[W|W] = \frac{1}{6} = 0.166$$

Therefore, the value of $P_W[W|W]$ has been verified, namely,

$$P_W[W|W] = 1 - P_W[W|B] = \frac{1}{6}$$

Up to this point, the transition probabilities related to the white jar are known, namely $P_W[W|B]$ and $P_W[W|W]$. Next, the transition probabilities related to the black jar must be determined, namely: $P_B[B|W]$ and $P_B[B|B]$. The sequence of observations indicates a total of five transitions (pairs of letters) from letter "B" to letter "W" ("W̶W̶BWBWBBWBWBBBW̶W̶"). The first letter "B" of the pair ("BW") appears eight times ("W̶WBWBWBBWBWBBBW̶W̶"):

$$P_B[B|W] = \frac{5}{8} = 0.625$$

In the case of the white jar, $P_W[W|W]$ has been determined from $P_W[W|B]$. The same reasoning can be applied for the next transition. Therefore, $P_B[B|B]$ is determined from $P_B[B|W]$, namely:

$$P_B[B|B] = 1 - P_B[B|W] = 0.375$$

These probabilistic determinations can stop here, but for the sake of illustrating the analysis, a continuation is made in order to check $P_B[B|B]$ the hard way. The great interest now lies in the number of times letter "B" appears in the sequence and the number of times "BB" pair (transitions from "B" to "B") appears in the sequence. Inside the sequence, the "BB" pair (representing the transition between "B" and "B") occurs three times ("W̶W̶BWBWBBWBWBBBWW"). The first letter "B" of the pair ("BB") appears eight times ("W̶WBWBWBBWBWBBBW̶W̶"), therefore:

$$P_B[B|B] = \frac{3}{8} = 0.375$$

Again, the long way of solving the problem showed that indeed:

$$P_B[B|B] = 1 - P_B[B|W] = \frac{3}{8} = 0.375$$

Now, it is perhaps evident that one known parameter for each jar allows for the reconstruction of the Markov diagram. To check the accuracy of the method, the probabilities corresponding to each jar must result in unity, namely $P_W[W|B] + P_W[W|W] = 1$ and $P_B[B|W] + P_B[B|B] = 1$:

$$P_W[W|B] + P_W[W|W] = \frac{5}{6} + \frac{1}{6} = 0.83 + 0.17 = 1$$

$$P_B[B|W] + P_B[B|B] = \frac{5}{8} + \frac{3}{8} = 0.625 + 0.375 = 1$$

The conditional probability between letters found above can be arranged in a simple table, called a transition probability matrix or a Markov matrix:

	W	B
W	0.166	0.833
B	0.625	0.375

The table format from above allows various observations on the data. (1) approximately 83.3% of the time the extractions will jump from the white to the black jar, and (2) almost 16.6% of the time the extractions will jump from the white jar to the white jar. From the black jar perspective, approximately 62.5% of the time the extractions will go from the black to the white jar and 37.5% of the time the extractions will go from the black jar to the black jar. The relationship between the Markov diagram and Markov matrix is shown in a graphic manner in Figure 2.1. Some additional cases can be studied in order to highlight the importance of what is *not* counted, namely: (1) the first and last state as well as (2) the transition between the first and the second state of the sequence (Table 2.1). As noted above, whether it is a "W" or a "B" letter, the counting of these observations is made from the second letter and ends at the penultimate letter in the sequence. Notice that in the examples shown in Table 2.1, the first and the last letter in the sequence are represented with a strikethrough line. The underline shows the letters that have been counted, and the orange color shows the transitions between letters that have been counted. Secondly, the only transition which is not counted is between the first and the second state (letter). Also, the transition which is not counted is represented with a strikethrough line through the first two letters (Table 2.1). However, the last letter in the sequence is counted in the case of transitions between letters (states).

2.4 The Simulation of a Two-State Markov Chain

So far, the transition probabilities have been determined from a sequence of observations provided, supposedly, by a system of two real jars. Following the above experiments, three questions may be of great interest, namely: Can a system of two jars be simulated on the computer? Could such a simulator mimic the draw behavior of the original system of jars? Can it produce a sequence of observations which resembles the sequence of observations of the original system of jars? A different number of new examples have been presented in a short manner (Table 2.1). For instance, consider the first sequence of observations (S_1) from Table 2.1, namely:

$$S_1 = \text{"WWBBWBWBBWBWBBBWWW"}$$

Sequence S_1 is provided by a system of two jars. Also, based on the sequence of observations (S_1) the transition matrix has been determined (Table 2.1). Probability values from the transition matrix dictate the proportions of balls from jars and vice versa. An overview of a transition matrix indicates that rows may represent the jars and the columns may represent the proportion of white and black balls in jars. Therefore, a new system of two jars can always be made to resemble the original system by complying with the data from the transition

Table 2.1 Seven examples that show the route from a sequence of observations to the transition matrix and the Markov diagram.

The observed sequence	Markov diagram	Transition probability matrix

S_1 = "WWBBWBWBBWBWBWBBBWW"

WWBBWBWBBWBWBWBBBWW = 5/6 = 0.83
WWBBWBWBBWBWBWBBBWW = 1/6 = 0.17
WWBBWBWBBWBWBWBBBWW = 5/9 = 0.56
WWBBWBWBBWBWBWBBBWW = 4/9 = 0.44

	to W	to B
From W	0.17	0.83
From B	0.56	0.44

S_2 = "WBBWBBBWBWWBBWBBBW"

WBBWBBBWBWWBBWBBBW = 4/5 = 0.8
WBBWBBBWBWWBBWBBBW = 1/5 = 0.2
WBBWBBBWBWWBBWBBBW = 5/10 = 0.5
WBBWBBBWBWWBBWBBBW = 5/10 = 0.5

	to W	to B
From W	0.2	0.8
From B	0.5	0.5

S_3 = "BWBWBWWBWBWBWBWWBBB"

BWBWBWWBWBWBWBWWBBB = 6/8 = 0.75
BWBWBWWBWBWBWBWWBBB = 2/8 = 0.25
BWBWBWWBWBWBWBWWBBB = 5/7 = 0.71
BWBWBWWBWBWBWBWWBBB = 2/7 = 0.29

	to W	to B
From W	0.25	0.75
From B	0.71	0.29

S_4 = "BBBBWWWWWWBWBWBWBWWW"

BBBBWWWWWWBWBWBWBWWW = 4/8 = 0.5
BBBBWWWWWWBWBWBWBWWW = 4/8 = 0.5
BBBBWWWWWWBWBWBWBWWW = 5/7 = 0.71
BBBBWWWWWWBWBWBWBWWW = 2/7 = 0.29

	to W	to B
From W	0.5	0.5
From B	0.71	0.29

(*continued*)

Table 2.1 (*Continued*)

The observed sequence	Markov diagram	Transition probability matrix
$S_5 = $ "BBWBWBWBBWBWBWWBW" BBWBWBWBBWBWBWWBW = 6/7 = 0.86 BBWBWBWBBWBWBWWBW = 1/7 = 0.14 BBWBWBWBBWBWBWWBW = 7/8 = 0.88 BBWBWBWBBWBWBWWBW = 1/8 = 0.12		From W: to W 0.14, to B 0.86 From B: to W 0.88, to B 0.12
$S_6 = $ "WWWWWWWBBBWBWBBBB" WWWWWWWBBBWBWBBBB = 3/8 = 0.38 WWWWWWWBBBWBWBBBB = 5/8 = 0.62 WWWWWWWBBBWBWBBBB = 2/7 = 0.29 WWWWWWWBBBWBWBBBB = 5/7 = 0.71		From W: to W 0.62, to B 0.38 From B: to W 0.29, to B 0.71
$S_7 = $ "WBWBWBWBBBBBWWBWB" WBWBWBWBBBBBWWBWB = 5/6 = 0.83 WBWBWBWBBBBBWWBWB = 1/6 = 0.17 WBWBWBWBBBBBWWBWB = 5/9 = 0.56 WBWBWBWBBBBBWWBWB = 4/9 = 0.44		From W: to W 0.17, to B 0.83 From B: to W 0.56, to B 0.44

matrix. Thus, the two new jars with the right proportions of white and black balls can mimic the observations from above (S_1). However, two issues must be considered for a computer simulation of these jars: (1) the representation of the jars in electronic format and (2) the simulation of draws.

2.4.1 The White Jar of the Original Machine

The sequence of observations (S_1) is generated by the initial machine (Table 2.1). The state diagram of the initial machine indicates that the first row from the transition matrix represents the probability of extracting a black or a white ball from the white jar. Therefore, the probability of extracting a white ball from the white jar is $P_W[W] = 0.17$ and the probability of extracting a black ball from the white jar is $P_W[B] = 1 - p$, namely 0.83. Similar to the Bernoulli example, these probabilities reflect the proportion of balls in the white jar. Therefore, inside the white jar, the white balls occupy 17% ($100 \times P_W[W] = 100 \times 0.17 = 17\%$) from the total and the black balls make up 83% ($100 \times P_W[B]$) of the total. The same reasoning can be applied for the black jar.

2.4.2 The Black Jar of the Original Machine

The state diagram of case S_1 also indicates that the second row from the transition matrix represents the probability of extracting a black or a white ball from the black jar (Table 2.1). Therefore, the probability of extracting a white ball from the black jar is $P_B[W] = 0.56$ and the probability of extracting a black ball from the black jar is $P_B[B] = 1 - p = 0.44$. Again, these probabilities reflect the proportion of balls in the black jar. Thus, the white balls in the black jar take 56% of the total and the black balls in the black jar make up what remains from the total, namely 44%.

2.4.3 The Representation of New Jars

In the original system, the precise number of balls in each jar is unknown; however, the proportions are known. In order to simulate a jar, a string of 10 letters (L_{tot}) can be used to represent 10 balls. Such a string may be composed of "W" and "B" letters that represent the proportions of white and black balls in each jar. The question that arises is: if a total of 10 letters (L_{tot}) are used, then what is the proportion of "W" and "B" letters? For instance, an example can be given by using the white jar. Ten letters represent 100% of the balls in the white jar representation. Intuitively, about two "W" letters represent approximately 17% of the total. Consequently, the other 83% is represented by eight letters "B" (Figure 2.3). However, in the new jar representation, a simple approach to actually

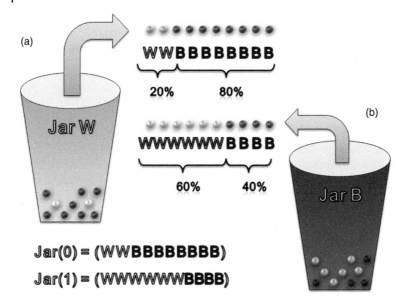

Jar(0) = (WWBBBBBBBB)
Jar(1) = (WWWWWWBBBB)

Figure 2.3 From jars to their representation in electronic format. (a) Linear representation of Jar "W". (b) Linear representation of Jar "B". Each jar contains 10 balls of different colors. The balls from each jar are linearly arranged by color in a string (the order does not matter). The string of balls is replaced with a string of letters. Each string of letters is then stored in a vector designated as "Jar".

calculate these proportions consists in the multiplication of probability values (belonging to the old jar) by the total number of letters (L_{tot}). A letter is an indivisible entity that must be represented by integers. In order to obtain integers, the values provided by the following formulas have to be rounded. Thus, the number of "W" letters inside the white jar representation (Jar$_W$) is:

$$\text{Jar}_W[W] = L_{tot} \times P_w[W] = 10 \times 0.17 = 1.7 = {\sim}2$$

The number of "B" letters inside the white jar (Jar$_W$) is:

$$\text{Jar}_W[B] = L_{tot} - \text{Jar}_W[W] = 10 - 1.7 = 8.3 = {\sim}8$$

Therefore, the sequence of letters for the white jar must include two letters "W" and eight letters "B"(Jar$_W$ = "WWBBBBBBBB"). Note also that the number of "W" and "B" letters from the white jar representation is equal to L_{tot}:

$$\text{Jar}_W[W] + \text{Jar}_W[B] = L_{tot}$$

By continuing this rationale, the number of "W" and "B" letters from the other jar can also be found, namely the black jar (Jar$_B$). Thus, the number of "W" letters inside the black jar is:

$$\text{Jar}_\text{B}[\text{W}] = L_\text{tot} \times P_\text{B}[\text{W}] = 10 \times 0.56 = 5.6 = \sim6$$

The proportion of "B" letters inside the black jar (Jar_B) is:

$$\text{Jar}_\text{B}[\text{B}] = L_\text{tot} - \text{Jar}_\text{B}[\text{W}] = 10 - 5.6 = 4.4 = \sim4$$

Therefore, the sequence of letters for the black jar representation must include six letters "W" and four letters "B" (Jar_B = "WWWWWWBBBB"). Again, note that the number of "W" and "B" letters from the black jar representation is equal to L_tot:

$$\text{Jar}_\text{B}[\text{W}] + \text{Jar}_\text{B}[\text{B}] = L_\text{tot}$$

Up to this point, Jar_W and Jar_B represent the rough copies of the original jar system that generated the S_1 sequence. The jar representations from the other cases ($S_2, \ldots, S_7$) follow the same rules and they are shown in Table 2.2.

2.4.4 Simulation of the System

One natural way to answer questions about Markov chains is to simulate them. For a computer simulation of this system, both jars (Jar_W and Jar_B) can be represented by a vector (Jar) with two components (Jar(0) and Jar(1)). Each component contains a sequence of 10 letters that represents the proportions of balls from a jar (Figure 2.3). Therefore, all the balls from the white jar (Jar_W) are represented by the first component, namely Jar(0), and all the balls from the black jar (Jar_B) are represented by Jar(1), the second component. In order to simulate a draw, a random integer is generated between 1 and 10 and that number represents the letter found at the ith position on a string of 10 letters. Thus, whether a ball is randomly drawn from a real jar or a letter is randomly chosen from a string of 10 letters, the process is exactly the same. For instance, the first component (Jar(0)) is representative for the white jar. Consider that a letter is randomly chosen from the sequence of the first component (Jar(0) = "WWBBBBBBBB"). The probability of randomly choosing a "W" is $P_\text{W}[\text{W}] = 2/10 = 0.2$ and the probability of randomly choosing letter "B" is $P_\text{W}[\text{B}] = 8/10 = 0.8$. The two systems will behave in a similar manner as long as the components of the vector ("Jar") contain the same proportions of "W" and "B" letters as the white and black balls in the original jars. Thus, a strategy for the representation of the jars on a computer has been described in detail. Nevertheless, one more aspect remains to be explained, namely the actual simulation of draws on a computer. Suppose that 17 draws must be simulated using the proportions given in the case of S_1 (Table 2.2). An algorithm implementation may use the vector components (i.e., Jar(0) and Jar(1)) precisely as they are declared in Table 2.2, second column. The implementation below mimics the initial system of jars (Supporting algorithm 1):

Table 2.2 The representation of the new jars for a series of cases.

Case	probabilities	The content of the jars	Transition matrix		
				to W	to B
S_1	$P_W[W] = 2/10 = 20\%$ $P_W[B] = 8/10 = 80\%$ $P_B[W] = 6/10 = 60\%$ $P_B[B] = 4/10 = 40\%$	Jar(0) = "WWBBBBBBBB" Jar(1) = "WWWWWWBBBB"	From W From B	0.17 0.56	0.83 0.44
				to W	to B
S_2	$P_W[W] = 2/10 = 20\%$ $P_W[B] = 8/10 = 80\%$ $P_B[W] = 5/10 = 50\%$ $P_B[B] = 5/10 = 50\%$	Jar(0) = "WWBBBBBBBB" Jar(1) = "WWWWWBBBBB"	From W From B	0.2 0.5	0.8 0.5
				to W	to B
S_3	$P_W[W] = 3/10 = 30\%$ $P_W[B] = 7/10 = 70\%$ $P_B[W] = 7/10 = 70\%$ $P_B[B] = 3/10 = 30\%$	Jar(0) = "WWWBBBBBBB" Jar(1) = "WWWWWWWBBB"	From W From B	0.25 0.71	0.75 0.29
				to W	to B
S_4	$P_W[W] = 5/10 = 50\%$ $P_W[B] = 5/10 = 50\%$ $P_B[W] = 7/10 = 70\%$ $P_B[B] = 3/10 = 30\%$	Jar(0) = "WWWWWBBBBB" Jar(1) = "WWWWWWWBBB"	From W From B	0.5 0.71	0.5 0.29
				to W	to B
S_5	$P_W[W] = 1/10 = 10\%$ $P_W[B] = 9/10 = 90\%$ $P_B[W] = 6/10 = 60\%$ $P_B[B] = 4/10 = 40\%$	Jar(0) = "WBBBBBBBBB" Jar(1) = "WWWWWWBBBB"	From W From B	0.14 0.88	0.86 0.12
				to W	to B
S_6	$P_W[W] = 4/10 = 40\%$ $P_W[B] = 6/10 = 60\%$ $P_B[W] = 3/10 = 30\%$ $P_B[B] = 7/10 = 70\%$	Jar(0) = "WWWWBBBBBB" Jar(1) = "WWWBBBBBBB"	From W From B	0.38 0.29	0.62 0.71
				to W	to B
S_7	$P_W[W] = 2/10 = 20\%$ $P_W[B] = 8/10 = 80\%$ $P_B[W] = 6/10 = 60\%$ $P_B[B] = 4/10 = 40\%$	Jar(0) = "WWBBBBBBBB" Jar(1) = "WWWWWWBBBB"	From W From B	0.17 0.56	0.83 0.44

```
Dim Jar(0 To 1) As String

Private Sub Form_Load()

draws = 17

Jar(0) = "WWBBBBBBBB"
Jar(1) = "WWWWWBBBBB"
```

```
a = Draw(0) ' Draws start from jar "W"
z = z & " Jar W[" & a & "],"


For i = 1 To draws


    If a = "W" Then
        a = Draw(0)
        z = z & " Jar W[" & a & "],"
    Else
        a = Draw(1)
        z = z & " Jar B[" & a & "],"
    End If


MsgBox z
Next i


End Sub


Function Draw(ByVal S As Integer) As String
    Randomize
    randomly_choose = Int(Rnd * Len(Jar(S)))
    ball = Mid(Jar(S), randomly_choose + 1, 1)
    Draw = ball
End Function
```

```
Output:
Jar W[B], Jar B[B], Jar B[B], Jar B[W], Jar W[B], Jar B[B], Jar
B[W], Jar W[B], Jar B[B], Jar B[B], Jar B[W], Jar W[B], Jar
B[B], Jar B[B], Jar B[W], Jar W[B], Jar B[B], Jar B[W],
```

Supporting algorithm 1. A two-state Markov chain simulator based on the proportions of letters. Two-letter sequences with predetermined proportions of "W" and "B" letters are used for the representation of two jars. The chance of a letter chosen at random from one of the two sequences is directly dictated by the proportions of "W" and "B" letters.

For a brief description, variable "*a*" stores a letter extracted at random from one of the components of the vector "jar" (from one of the two jars). This letter is returned to variable "*a*" by function "Draw". Before being called, function "Draw" may receive an integer variable "*S*" which specifies the component used (the jar used). Each component of the vector "jar" stores one string of 10 letters. In order to simulate a draw, function "Draw" generates a random integer between 1 and 10 that represents the letter found at the ith position on a string of 10 letters. Note that vector "Jar" is declared globally, so it can be read both by the main function and other functions. Next, the white jar is the first jar (state) from which draws begin. In our particular case the white jar is represented by the vector component "Jar(0)". This first draw is initiated by a call to function "Draw". Therefore, the integer variable "*S*" takes zero as the first value. If "Draw" function returns letter "W" to variable "*a*", then the next draw is made from the white jar (represented by component "Jar(0)"). If the function returns letter "B" to variable "*a*", then the next draw is made from the black jar (represented by component "Jar(1)"). This process continues until the 17th draw. Of course, if only 10 letters are used the new system will include rounded probabilities. For instance, two letters from a total of 10 letters is 0.2 instead of 0.17, and 8 letters from a total of 10 letters is 0.8 instead of 0.83 as it was originally for the S_1 case. Obviously, to represent the balls in a jar and comply with their relevant percentages, a total of 100 letters may be used instead of 10 letters. Since the only change compared to the previous example is $L_{tot} = 100$, the following can be written:

$$\text{Jar}_W[W] = L_{tot} \times P_w[W] = 100 \times 0.17 = 17$$
$$\text{Jar}_W[B] = L_{tot} - \text{Jar}_W[W] = 100 - 17 = 83$$
$$\text{Jar}_B[W] = L_{tot} \times P_B[W] = 100 \times 0.56 = 56$$
$$\text{Jar}_B[B] = L_{tot} - \text{Jar}_B[W] = 100 - 56 = 44$$

Again note that the number of "W" and "B" letters from each jar representation is equal to L_{tot}:

$$\text{Jar}_B[W] + \text{Jar}_B[B] = L_{tot}$$
$$\text{Jar}_W[W] + \text{Jar}_W[B] = L_{tot}$$

It is relatively time consuming to manually create the proportions of "W" and "B" letters in a string of 100 letters. Thus, these jar representations can be constructed with a specialized function. Therefore, the implementation below brings a new function ("Fill Jar") that is responsible for building the strings that are stored in each of the vector components:

```
Dim Jar(0 To 1) As String

Private Sub Form_Load()

Call Fill_Jar(0, 0.2) 'W
Call Fill_Jar(1, 0.6) 'B

draws = 17

a = Draw(0) ' Draws start from jar "W"
z = z & " Jar W[" & a & "],"
```

```
For i = 1 To draws
    If a = "W" Then
        a = Draw(0)
        z = z & " Jar W[" & a & "],"
    Else
        a = Draw(1)
        z = z & " Jar B[" & a & "],"
    End If
MsgBox z
Next i

End Sub

Function Draw(ByVal S As Variant) As String
    Randomize
    randomly_choose = Int(Rnd * Len(Jar(S)))
    ball = Mid(Jar(S), randomly_choose + 1, 1)
    Draw = ball
End Function

Function Fill_Jar(ByVal S As Integer, ByVal p As Variant)
Balls_W = Int(100 * p)
Balls_B = 100 - Balls_W

For i = 1 To Balls_W
    Jar(S) = Jar(S) & "W"
Next i

For i = 1 To Balls_B
    Jar(S) = Jar(S) & "B"
Next i

End Function
```

```
Output:
Jar W[B], Jar B[W], Jar W[B], Jar B[W], Jar W[B], Jar B[B], Jar
B[W], Jar W[B], Jar B[B], Jar B[B], Jar B[W], Jar W[B], Jar
B[B], Jar B[B], Jar B[W], Jar W[B], Jar B[B], Jar B[W]
```

Supporting algorithm 2. A two states Markov chain simulator based on probability values. The probability values present inside the transition matrix are directly used for an automatic generation of the letter combination that make up the representation of the jars. Thus, the two letter sequences have a calculated proportion of "W" and "B" letters. The chance of a letter chosen at random from one of the two sequences is directly dictated by the proportions of "W" and "B" letters.

Function "Fill_Jar" may receive two parameters, namely "S" and "p". The first parameter which can be received by function "Fill_Jar" is an integer variable "S" which specifies the component to be filled with suitable proportions of "W" and "B" letters. Therefore, because only two states are involved (two jars) the integer variable "S" takes 0 or 1 as possible values. The second parameter is represented by the probability of the letter "W" for the string of that component. For instance, if $S = 0$ then the value of $P_W[W]$ is used, otherwise the value of $P_B[W]$ is used. Inside the function, $P_W[B]$ or $P_B[B]$ are determined by $1 - p$, namely $P_W[B] = 1 - P_W[W]$ and $P_B[B] = 1 - P_B[W]$. Once the "Fill_Jar" function was called for each component of the vector "Jar", the events remain similar to the previous example.

3

Building the Stochastic Matrix

3.1 Introduction

Black and white jars have been used to represent the states of an abstract machine, whereas white or black balls held the role of the observable. In nature, there are many observable events which may be modeled as Markov chains. Weather-related events are the most classic example [18–25]. This chapter describes the method by which a transition matrix of a two-state Markov chain can be obtained from a sequence of events or from some relative observations expressed in percentages. Many concepts regarding the structure and properties of a transition matrix have been previously discussed; nevertheless, the following sections cover both the theory and the implementation.

3.2 Building a Stochastic Matrix from Events

Transition probabilities can be retrieved from known events. For instance, mutually exclusive events can be very useful for modeling a discrete-time Markov chain. Some examples of mutually exclusive events per unit of time (seconds, minutes, days, or years for instance) can be weather-related, such as sunny (noted as S) or rainy (noted as R). A sunny day and a rainy day are observations that may be considered mutually exclusive. For instance, the weather observations on a previous period of 15 days can be modeled by using two states, namely S or R (Figure 3.1). Thus, daily observations of the weather can be written in a string of letters such as: "SRRSRSRRS…" Suppose that the sequence (E) contains a total of 15 observations (i.e., E = "SRRSRSRRSRSRRSS"). Thus, transitions are represented by the pairs of letters present inside the series of observations (E). Onward, two values are of interest for calculating transition probabilities. First, the number of pairs is counted. Second, the first letter of the pair is counted independently. In order to compute the transition probabilities, the number of pairs is divided by the number of single letters. Thus, the

Markov Chains: From Theory to Implementation and Experimentation, First Edition. Paul A. Gagniuc.
© 2017 John Wiley & Sons, Inc. Published 2017 by John Wiley & Sons, Inc.
Companion website: www.wiley.com/go/gagniuc/markovchains

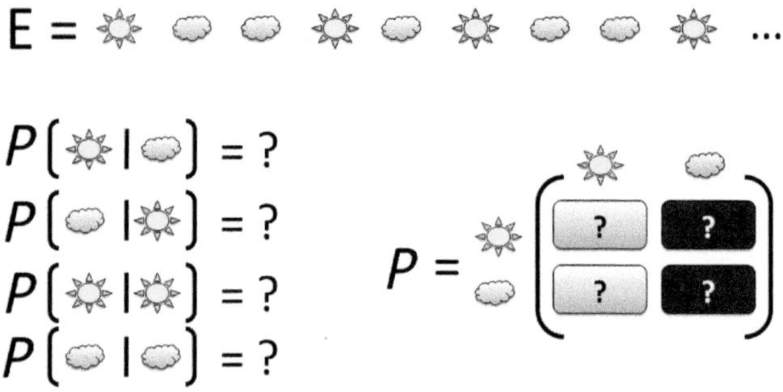

$$E = \text{☀} \; \text{☁} \; \text{☁} \; \text{☀} \; \text{☁} \; \text{☀} \; \text{☁} \; \text{☁} \; \text{☀} \; \ldots$$

$$P\left(\text{☀} \mid \text{☁}\right) = ?$$
$$P\left(\text{☁} \mid \text{☀}\right) = ?$$
$$P\left(\text{☀} \mid \text{☀}\right) = ?$$
$$P\left(\text{☁} \mid \text{☁}\right) = ?$$

$$P = \begin{pmatrix} ? & ? \\ ? & ? \end{pmatrix}$$

Figure 3.1 An overall view for building a transition matrix from known events. Each icon in the sequence of E represents an observation made for an entire day. The transition matrix is schematically illustrated on the right side of the figure. The left side of the figure illustrates the conditional probability functions associated with the matrix.

following conditional probability functions can be written as follows:

$$P(R|S) = \frac{\text{Count(SR)}}{\text{Count(S)}}$$

$$P(S|S) = \frac{\text{Count(SS)}}{\text{Count(S)}}$$

$$P(S|R) = \frac{\text{Count(RS)}}{\text{Count(R)}}$$

$$P(R|R) = \frac{\text{Count(RR)}}{\text{Count(R)}}$$

where, for instance, the meaning of the conditional probability function $P(R|S)$ is: the probability of event "R" given event "S" already occurred. The conditional probability function can be perhaps less intuitive when the word "transition" is taken into context. Note that $P(R|S)$ and $P[S|R]$ have the same meaning. Both notations signify a transition from "S" to "R" (see Glossary). However, a more straightforward notation can encapsulate the above four expressions into a single formula, namely:

$$T_{a \to b} = \frac{\text{Count}(D_{a \to b})}{\text{Count}(N_a)}$$

where $T_{a \to b}$ is the transition probability from an event that represents a, to an event that represents b. $D_{a \to b}$ represents the number of transitions from a to b, and N_a represents the number of times a appears in the sequence. *For example*, consider the transition probability from "S" (sunny) to "R" (rainy) in the sequence "SRRSRSRRSRSRRSS". In this case, a is represented by the letter "S" and b is represented by the letter "R" (Figures 3.2a and 3.2b). Furthermore, $D_{a \to b}$ counts how many times "SR" appears in the sequence

(a)

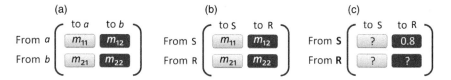

(b)

(c)

Figure 3.2 Building a right stochastic matrix. (a) Transition probabilities from events a to events b. (b) Events a are represented by letters from which the transition is made and events b are represented by letters to which the transition is made. (c) Following the examples above, by calculating $T_{S \to R}$ the transition probability from "S" to "R" has been found.

"S̶R̶RSRSRRSRSRRSS", and N_a counts how many times "S" appears in the sequence "S̶RRSRSRRSRSRRS̶S̶". Therefore, the formula becomes:

$$T_{S \to R} = \frac{\text{Count}(D_{S \to R})}{\text{Count}(N_S)} = \frac{4}{5} = 0.8$$

where $D_{S \to R}$ appears four times and N_S appears five times. Four divided by five provides a value of 0.8. Thus, the first value in the transition matrix has been found (Figure 3.2c). The above formula can be considered further in the same manner. The other transition probability values can be found by replacing a and b with relevant observations as follows:

$$T_{S \to R} = \frac{\text{Count}(D_{S \to R})}{\text{Count}(N_S)}$$

$$T_{S \to S} = \frac{\text{Count}(D_{S \to S})}{\text{Count}(N_S)}$$

$$T_{R \to S} = \frac{\text{Count}(D_{R \to S})}{\text{Count}(N_R)}$$

$$T_{R \to R} = \frac{\text{Count}(D_{R \to R})}{\text{Count}(N_R)}$$

First, the pairs of letters $(D_{a \to b})$ are counted for each of the four types of transitions. The sequence of events (E) contains four pairs of "SR" ("S̶R̶RSRSR̶R̶SRSRRSS"), five pairs of "RS" ("S̶R̶R̶S̶RSRRSRSRRSS"), three pairs of "RR" ("SRRSRSR̶R̶SRSRRSS") and one pair of "SS" ("SRRSRSRRSRSRRS̶S̶"). A counting table can be formed based on $D_{a \to b}$ pairs of letters. In this table, the rows represent the first letter from the pairs and the columns represent the second letter from the pairs:

	S	R
S	$D_{S \to S}$	$D_{S \to R}$
R	$D_{R \to S}$	$D_{R \to R}$

The coordinates for the count values have been established in the above table. Thus, since $D_{S \to R} = 4$, $D_{R \to S} = 5$, $D_{R \to R} = 3$, $D_{S \to S} = 1$, the above table can be filled with the following values:

	S	R
S	1	4
R	5	3

For the final step, a second parameter is of interest, namely N_a. Notice the total number of appearances of each state (N_a) in the sequence. Sunny ("S") occurs five times ("SRRSRSRRSRSRRSS") while rainy ("R") occurs eight times ("SRRSRSRRSRSRRSS"). Thus, $N_S = 5$ and $N_R = 8$. By replacing the values in the above formulas, the following transition probabilities can be obtained:

$$T_{S \to R} = \frac{4}{5} = 0.8$$

$$T_{S \to S} = \frac{1}{5} = 0.2$$

$$T_{R \to S} = \frac{5}{8} = 0.625$$

$$T_{R \to R} = \frac{3}{8} = 0.375$$

Note: The first and last letters from the sequence of observations are not taken into account when calculating the transition probabilities ("SRRSRSRRSRSRRSS"). Consider an experiment made with a system of two jars. Before the first draw, the transition is made from *nowhere*. In other words, the first state is *chosen* by the experimenter and is not a part of the stochastic process. After the last draw, the transition is made to *nowhere*. Therefore, the last state to which the transition should be made is unknown. Also, the first transition (pair of letters) in the sequence cannot be counted since it is made from the first letter (state) that has been *chosen* by the experimenter ("SRRSRSRRSRSRRSS"). Forward, the count table from above is converted to a 2 × 2 transition probability matrix. This is obtained by dividing $D_{a \to b}$ integers to N_a pieces:

	S		R	
S	$T_{S \to S} = \dfrac{\text{Count}(D_{S \to S})}{\text{Count}(N_S)}$		$T_{S \to R} = \dfrac{\text{Count}(D_{S \to R})}{\text{Count}(N_S)}$	
R	$T_{R \to S} = \dfrac{\text{Count}(D_{R \to S})}{\text{Count}(N_R)}$		$T_{R \to R} = \dfrac{\text{Count}(D_{R \to R})}{\text{Count}(N_R)}$	

Table 3.1 From a weather-related sequence (E) of observations to the Markov diagram and the transition matrix.

The observed sequence	Markov diagram	Transition probability matrix
E = "SRRSRSRRSRSRRSS" S̶RRSRSRRSRSRRS̶S̶ = 4/5 = 0.8 S̶RRSRSRRSRSRRSS = 1/5 = 0.2 S̶RRSRSRRSRSRRSS = 5/8 = 0.625 S̶RRSRSRRSRSRRSS = 3/8 = 0.375		

which results in a right stochastic matrix (Figure 2.2a):

	S	R
S	0.2	0.8
R	0.625	0.375

Notice that the sum of probability values from each row equals 1. Up to this point, a transition matrix has been obtained from a weather-related sequence of observations. As with the jars and balls examples, a summary of those described above can be made (Table 3.1). Notice that in Table 3.1, the first and the last letter in the sequence are represented with a strikethrough line. The underline shows the single letters (N_a) that have been counted, and the orange color shows the pairs of letters ($D_{a \to b}$) that have been counted. The only transition (pair of letters) which is not counted resides between the first and the second letter (state) in the sequence. Also in Table 3.1, the transition which is not counted is represented with a strikethrough line through the first two letters. However, the last pair of letters in the sequence is counted (see Note above). When the implementation is made, the transition matrix has a classical coordinate system. This coordinate system denotes the intersection of the row (i) with the column (j), thereby indicating a specific element inside the matrix. Thus, the matrix elements are denoted as $m_{i,j}$. For the example shown above, the elements of the matrix are written as follows:

	S	R
S	m_{11}	m_{12}
R	m_{21}	m_{22}

In relation to the above example, the coordinate system may have the following equivalence:

$$m_{11} = T_{S \to S}$$
$$m_{12} = T_{S \to R}$$
$$m_{21} = T_{R \to S}$$
$$m_{22} = T_{R \to R}$$

Therefore, iterations can be made when the m_{ij} reference system is applied for the elements of the transition matrix. The implementation below computes the transition probabilities based on the sequence of observations "SRRSRSRRSRSRRSS" and generates a transition matrix:

```
Dim M(1 To 2, 1 To 2) As String

Private Sub Form_Load()
Call ExtractProb("SRRSRSRRSRSRRSS")
End Sub

Function ExtractProb(ByVal s As String)
Eb = "S"
Es = "R"

For i = 1 To 2
    For j = 1 To 2
      M(i, j) = 0
    Next j
Next i

TB = 0
TS = 0

For i = 2 To Len(s) - 1
        DI1 = Mid(s, i, 1)
        DI2 = Mid(s, i + 1, 1)

        If DI1 = Eb Then R = 1
        If DI1 = Es Then R = 2
        If DI2 = Eb Then c = 1
        If DI2 = Es Then c = 2

        M(R, c) = Val(M(R, c)) + 1
```

```
        If DI1 = Eb Then TB = TB + 1
        If DI1 = Es Then TS = TS + 1
Next i

MsgBox DrowMatrix(2, 2, M, "(C)", "Count:")

For i = 1 To 2
    For j = 1 To 2
        If i = 1 Then M(i, j) = Val(M(i, j)) / TB
        If i = 2 Then M(i, j) = Val(M(i, j)) / TS
    Next j
Next i

MsgBox DrowMatrix(2, 2, M, "(P)", "Transition matrix M:")
End Function

Function DrowMatrix(ib, jb, ByVal M As Variant, ByVal model
As String, ByVal msg As String) As String

Eb = "S"
Es = "R"

y = "|___|___|___|"
ct = ct & vbCrLf & "_____"
ct = ct & vbCrLf & "| " & model & " |   " & Eb & "   |   " & Es
& "  | "
ct = ct & vbCrLf & y & vbCrLf

For i = 1 To ib
    For j = 1 To jb

    v = Round(M(i, j), 2)

        If Len(v) = 0 Then u = "|        "
        If Len(v) = 1 Then u = "|       "
        If Len(v) = 2 Then u = "|      "
        If Len(v) = 3 Then u = "|     "
        If Len(v) = 4 Then u = "|    "
        If Len(v) = 5 Then u = "||"

        If j = jb Then o = "||" Else o = ""
        If j = 1 And i = 1 Then ct = ct & "|   " & Eb & "   "
        If j = 1 And i = 2 Then ct = ct & "|   " & Es & "   "
```

```
              ct = ct & u & v & o
        Next j
   ct = ct & vbCrLf & y & vbCrLf
   Next i


   DrowMatrix = msg & " M[" & Val(jb) & "," & Val(ib) & "]" &
   vbCrLf & ct & vbCrLf
   End Function
```

```
Output:
Count: M[2,2]
|(C)| S | R |
|___|___|___|
| S | 1 | 4 |
|___|___|___|
| R | 5 | 3 |
|___|___|___|

Transition matrix M: M[2,2]
_____
|(P)| S | R |
|___|___|___|
| S |0.2|0.8|
|___|___|___|
| R |0.6|0.4|
|___|___|___|
```

Supporting algorithm 3: The conversion of a sequence of observations to a transition matrix.
A 2 × 2 matrix is used for counting all the combinations of pairs of letters ($D_{a \to b}$) in the sequence ($D_{a \to b}$ is represented by joining two string variables, namely DI1 and DI2). In parallel, the first letter of each pair (N_a) is counted inside the sequence (N_a is represented by variable DI1). Next, the values from each element of the 2 × 2 matrix are divided by the number of first letters found in the sequence. Depending on the type of values (counts or probability values) stored inside, the same matrix is then shown twice in a graphical format.

3.3 Building a Stochastic Matrix from Percentages

The probability of events can also be deduced from relative observations. Such observations may consist of values provided in percentages. A percentage value (x) can be transformed into a probability value (P) by dividing x to 100 ($P = x/100$). Consider a world in which there are only two types of mutually exclusive weather phenomena: sunny and rainy. In this example, a sunny day is noticed 90% of the time for a certain period. This percentage is indicated by x. Therefore, the value of $P(\text{sunny}) = x/100 = 90/100 = 0.9$. The two mutually

exclusive weather phenomena impose a probability vector v with two components. The first component of vector v contains the probability value for a sunny
day, whereas the second component contains the probability value for a rainy
day $v = [P(\text{sunny})\ P(\text{rainy})]$. Vector v contains one known value ($P(\text{sunny})$) and
an unknown value ($P(\text{rainy})$). The sum of the two components in any given situation must be equal to 1. Thus, since $P(\text{sunny}) = 0.9$, the unknown probability
of $P(\text{rainy})$ is:

$$P(\text{sunny}) + P(\text{rainy}) = 1$$

$$P(\text{rainy}) = 1 - P(\text{sunny})$$

$$P(\text{rainy}) = 1 - 0.9$$

$$P(\text{rainy}) = 0.1$$

Moreover, a two-state Markov chain can be represented by *a right stochastic matrix made from two probability vectors positioned one over the other.*
Each probability vector includes two values (components). Taken *separately,*
the two probability vectors can be regarded as representatives for *independent variables.* Therefore, if the value of a component ($P[Y]$) is unknown and
the other component ($P[X]$) is known, then the unknown component ($P[Y]$)
can be deduced by $P[Y] = 1 - P[X]$ (unknown = 1 − known). Thus, a simple
condition is required for filling the unknown values of a transition matrix. At
least one value (component) from each probability vector must be known. For
instance, the transition probabilities of a two-state Markov chain can be easily determined if the observations are provided in percentages. As an example,
the matrix P represents the weather model in which a sunny day (noted as "S")
is 90% likely to be followed by another sunny day, and a rainy day (noted as
"R") is 50% likely to be followed by another rainy day (Figure 3.3a). Thus, the
transition probability from "S" to "S" (noted $P[S|S]$) is 0.9 and the transition
probability from "R" to "R" (noted $P[R|R]$) is 0.5 (Figure 3.3a). Since there is
one known transition probability from each probability vector, the unknown
values can be deduced, namely $P[S|R]$ (alternative notation $T_{S\rightarrow R}$) and $P[R|S]$
(alternative notation $T_{R\rightarrow S}$):

$$P[S|R] = 1 - P[S|S] = 1 - 0.9 = 0.1$$
$$P[R|S] = 1 - P[R|R] = 1 - 0.5 = 0.5$$

Note that $P[S|R]$ has the same meaning as a conditional probability func-
tion, noted $P(R|S)$, where the conditional probability function $P(R|S)$ is the
probability of event "R" given event "S" already occurred. In the alternative

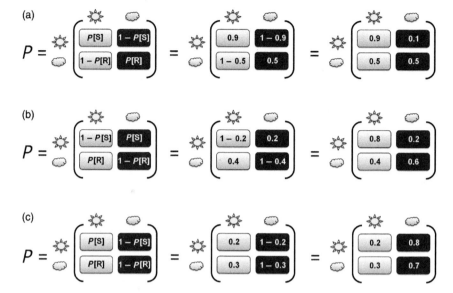

Figure 3.3 Filling the unknowns inside the transition matrix. A right stochastic matrix is made from two probability vectors positioned one over the other. Inside the matrices, these probability vectors are viewed independently of each other. (a) Known transitions from a sunny day to a sunny day ($P[S|S] = 0.9$) and from a rainy day to a rainy day ($P[R|R] = 0.5$). (b) Known transitions from a sunny day to a rainy day ($P[S|R] = 0.2$) and from a rainy day to a sunny day ($P[R|S] = 0.4$). (c) Known transitions from a sunny day to a sunny day ($P[S|S] = 0.2$) and from a rainy day to a sunny day ($P[R|S] = 0.3$).

notation, a conditional probability function $P(R|S)$ is written as $P[S|R]$ or $T_{S \to R}$, which means the transition from event "**S**" to event "**R**". Also, note that $P[R|S]$ is equivalent to a conditional probability function $P(S|R)$, where $P(S|R)$ is the probability of event "**S**" given event "**R**" already occurred. Nevertheless, both notations signify a transition from "**S**" to "**R**" (see Glossary). If the alternative notation is preferred, then the above expressions can be written also as:

$$T_{S \to R} = 1 - T_{S \to S} = 1 - 0.9 = 0.1$$
$$T_{R \to S} = 1 - T_{R \to R} = 1 - 0.5 = 0.5$$

Consider another example in which a sunny day is 20% likely to be followed by a rainy day and a rainy day is 40% likely to be followed by a sunny day (Figure 3.3b). Thus, the transition probability from "**S**" to "**R**" (noted $P[S|R]$) is 0.2 and the transition probability from "**R**" to "**S**" (noted $P[R|S]$) is 0.4 (Figure 3.3b). Since one transition probability is known from each probability vector, the unknown values can be deduced, namely $P[S|S]$ and $P[R|R]$:

$$P[S|S] = 1 - P[S|R] = 1 - 0.2 = 0.8$$
$$P[R|R] = 1 - P[R|S] = 1 - 0.4 = 0.6$$

Also, another example can be taken in which a sunny day is 20% likely to be followed by a sunny day ($P[S|S] = 0.2$) and a rainy day is 30% likely to be followed by a sunny day ($P[R|S] = 0.3$). At least one transition probability for each probability vector is known. Thus, the unknown values can be deduced, namely $P[S|R]$ and $P[R|R]$ (Figure 3.3c):

$$P[S|R] = 1 - P[S|S] = 1 - 0.2 = 0.8$$
$$P[R|R] = 1 - P[R|S] = 1 - 0.3 = 0.7$$

However, if the values of $P[S|S]$ (alternative notation $T_{S \rightarrow S}$) and $P[S|R]$ (alternative notation $T_{S \rightarrow R}$) are known, then both components of the second vector ($P[R|S]$ and $P[R|R]$) will be unknown and therefore impossible to deduce. Vice versa, when the values of $P[R|S]$ and $P[R|R]$ are known, the values of $P[S|S]$ and $P[S|R]$ are impossible to deduce.

4

Predictions Using Two-State Markov Chains

4.1 Introduction

First, the chapter describes the method of prediction by using the transition matrix of a two-state Markov chain. Second, the chapter describes the notion of steady-state vector in which the Markov chain reaches the limit of prediction. Also, the chapter brings to light the long-run distribution of a Markov chain which shows the convergence toward the steady-state vector, which represents the natural path of the machine to equilibrium. The algorithm implementations are shown separately for each step.

4.2 Performing the Predictions by Using the Stochastic Matrix

A parallel was made between two systems. The first system consisted of balls and jars (black or white balls), and the second system consisted of weather-related observations (sunny or rainy). So far, a transition matrix has been obtained from a sequence of events. But what can be done with the transition matrix? How can the following events be predicted? Can the weather be predicted for tomorrow and the day after tomorrow? Consider the sequence of events E = "SRRSRSRRSRSRRS" and the final letter in this sequence which represents the current observation (S). Thus, inside the sequence (E), tomorrow is temporarily represented by a question mark ("?"): E = "SRRSRSRRSRSRRSS?". The Markov diagram for the sequence E is shown in Figure 4.1. A right stochastic matrix (P) has already been built for the sequence of events (E). Previously, the calculated values for the transition probabilities have been (Figure 4.1):

$$P = \begin{bmatrix} P[S|S] & P[S|R] \\ P[R|S] & P[R|R] \end{bmatrix} = \begin{bmatrix} 0.2 & 0.8 \\ 0.625 & 0.375 \end{bmatrix}$$

Markov Chains: From Theory to Implementation and Experimentation, First Edition. Paul A. Gagniuc.
© 2017 John Wiley & Sons, Inc. Published 2017 by John Wiley & Sons, Inc.
Companion website: www.wiley.com/go/gagniuc/markovchains

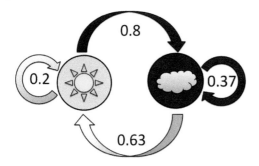

Figure 4.1 The Markov diagram for sequence E = "SRRSRSRRSRSRRSS".

where a transition probability is written as $P[\text{from}|\text{to}]$. To move forward, an initial state vector may be constructed ($v^{(0)}$). The first component of the new vector contains the probability value for a sunny day ($P[S]$), whereas the second component contains the probability value for a rainy day ($P[R]$):

$$v^{(0)} = [\, P[S] \quad P[R] \,]$$

Since the current observation is sunny (<u>S</u>), the probability of sunny is 1 and, consequently, the probability of rainy is $1 - p$, namely 0:

$$v^{(0)} = [\, 1 \quad 0 \,]$$

The coordinates of the transition matrix are replaced with probability values:

$$P = \begin{bmatrix} m_{11} & m_{21} \\ m_{12} & m_{22} \end{bmatrix} = \begin{bmatrix} 0.2 & 0.8 \\ 0.625 & 0.375 \end{bmatrix}$$

Notice that $[m_{11}\ m_{12}]$ and $[m_{21}\ m_{22}]$ are the two probability vectors that make up the transition matrix (P). Next, the coordinates of the initial state vector $v^{(0)}$ are written as:

$$v^{(0)} = [\, x_0 \quad y_0 \,] = [\, 1 \quad 0 \,]$$

where x_0 is the probability of a sunny day and y_0 is the probability of a rainy day, both inferred from the current observation (today). A new vector ($v^{(1)}$) can be obtained by using the transition matrix (P) and the initial state vector ($v^{(0)}$) from above:

$$v^{(1)} = [\, x_1 \quad y_1 \,] = [\, ? \quad ? \,]$$

In order to calculate the components of the new vector, some rules must be considered (Figure 4.2). These rules show the manner in which the initial state vector ($v^{(0)}$) interacts with the transition matrix (P). Therefore, the probability

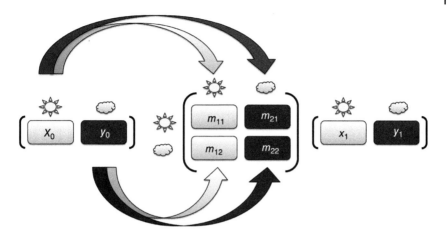

Figure 4.2 The prediction rules. The initial state vector ($v^{(0)}$) and the transition matrix (P) are multiplied. The first component (x_1) of the new vector ($v^{(1)}$) is calculated as $x_1 = (x_0 \times m_{11})$ + ($y_0 \times m_{12}$). The second component (y_1) of the new vector ($v^{(1)}$) is calculated as $y_1 = (x_0 \times m_{21})$ + ($y_0 \times m_{22}$).

of a sunny day tomorrow is represented by the first component (x_1) of the new vector ($v^{(1)}$), which is calculated as follows (Figure 4.2):

$$x_1 = (x_0 \times m_{11}) + (y_0 \times m_{12})$$

The above expression can also be written as:

$$x_1 = (P[S] \times P[S|S]) + (P[R] \times P[S|R])$$

A probability vector p is composed of v_n components arranged vertically or horizontally:

$$v = [p_1 \quad p_2 \quad \cdots \quad p_n]$$

$$\text{or } v = \begin{bmatrix} p_1 \\ p_2 \\ \vdots \\ p_n \end{bmatrix}$$

The values of the components of the probability vector v sum to 1:

$$\sum_{i=1}^{n} v_i = 1$$

Each individual component has a probability value between 0 and 1:

$$0 \leq p_i \leq 1$$

The probability of a rainy day tomorrow is represented by the second component (y_1) of the new vector ($v^{(1)}$), which is calculated in a similar manner (Figure 4.2):

$$y_1 = (x_0 \times m_{21}) + (y_0 \times m_{22})$$

The above expression can also be written as:

$$y_1 = (P[S] \times P[R|S]) + (P[R] \times P[R|R])$$

If the probability vector ($v^{(0)}$) and the transition matrix (P) are placed in context, then the next vector ($v^{(1)}$) can be obtained:

$$v^{(0)} \times P = v^{(1)}$$

$$[x_0 \, y_0] \begin{bmatrix} m_{11} & m_{21} \\ m_{12} & m_{22} \end{bmatrix} = [x_1 \quad y_1]$$

$$[1 \quad 0] \begin{bmatrix} 0.2 & 0.8 \\ 0.625 & 0.375 \end{bmatrix} = [x_1 \quad y_1]$$

The first component (x_1) of the new probability vector is evaluated:

$$x_1 = (x_0 \times m_{11}) + (y_0 \times m_{12})$$
$$x_1 = (1 \times 0.2) + (0 \times 0.625)$$
$$x_1 = (0.2) + (0)$$
$$x_1 = 0.2$$

The first probability value was found and can be inserted in context:

$$[1 \quad 0] \begin{bmatrix} 0.2 & 0.8 \\ 0.625 & 0.375 \end{bmatrix} = [0.2 \quad y_1]$$

When two states are used, a "shortcut" can be immediately observed. The components of any probability vector must sum to one. If a component of the new vector is found, the other component can be inferred by $1 - P$, and in this case:

$$y_1 = 1 - P[x_1] = 1 - 0.2 = 0.8$$

However, for an assessment in the manner discussed above, the second component (y_1) of the new probability vector must be evaluated without any shortcuts:

$$y_1 = (x_0 \times m_{21}) + (y_0 \times m_{22})$$
$$y_1 = (1 \times 0.8) + (0 \times 0.375)$$
$$y_1 = (0.8) + (0)$$
$$y_1 = 0.8$$

The second probability value was found and can be inserted in context:

$$[1 \quad 0] \begin{bmatrix} 0.2 & 0.8 \\ 0.625 & 0.375 \end{bmatrix} = [0.2 \quad 0.8]$$

In the next phase, the resulting vector is evaluated by determining which of its components contains a higher value. The probability of a sunny day for tomorrow is represented by the vector component x_1 and the probability of a rainy day for tomorrow is represented by the vector component y_1. Component x_1 has a value of 0.2 and component y_1 has a value of 0.8.

$$\text{Day}(v^{(1)}) = \begin{cases} R, & x_1 < y_1 \\ S, & x_1 \geq y_1 \end{cases}$$

Therefore, since y_1 has a higher value than x_1 ($x_1 < y_1$), the prediction for tomorrow indicates that it is more likely to be a rainy day (Day ($v^{(1)}$) = "R"). Following the example from the beginning, the question mark ("?") from the sequence "SRRSRSRRSRSRRSS?" can be replaced with an "R" (E = "SRRSRSRRSRSRRSSR"). Up to this point, the prediction for the first day has been made. The initial state vector ($v^{(0)}$) and the transition matrix (P) have been multiplied in order to obtain the new probability vector $v^{(1)}$ (Figure 4.3):

$$v^{(0)} \times P = v^{(1)}$$

where:

$$v^{(0)} = [\, x_0 \quad y_0 \,]$$
$$P = \begin{bmatrix} m_{11} & m_{21} \\ m_{12} & m_{22} \end{bmatrix}$$
$$v^{(1)} = [\, x_1 \quad y_1 \,]$$

So far the weather for tomorrow has been successfully predicted. Nevertheless, two important questions arise: (1) Can a weather prediction be made for several days? (2) What is the limit of this prediction? To answer the first question, the new sequence E ("SRRSRSRRSRSRRSSR") should be considered. In order to obtain the probability vector representative for day 2 ($v^{(2)}$), the resulting vector ($v^{(1)}$) is again multiplied by the transition matrix (P). This chain rule is repeated depending on how many days the prediction is made. Thus, the following expressions can be written for a certain number of days (k):

$$v^{(0)} \times P = v^{(1)}$$
$$v^{(1)} \times P = v^{(2)}$$
$$v^{(2)} \times P = v^{(3)}$$
$$v^{(3)} \times P = v^{(4)}$$
$$v^{(4)} \times P = v^{(5)}$$
$$v^{(5)} \times P = v^{(6)}$$
$$\vdots$$
$$v^{(k-1)} \times P = v^{(k)}$$

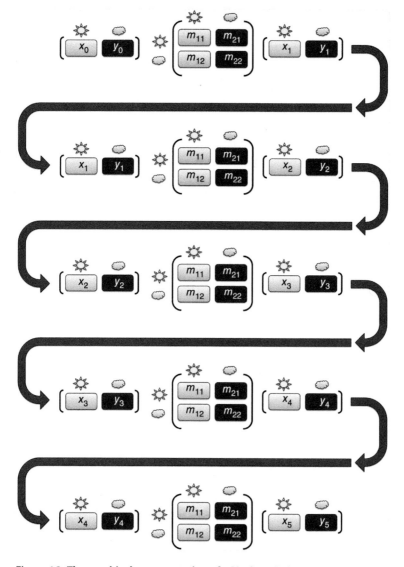

Figure 4.3 The graphical representation of a Markov chain.

A detailed example can be given for a weather prediction on the next 5 days in a row (Figure 4.3). The sequence (E) has been shown at the beginning of the chapter and it is further used for illustration. The prediction for day 1 was previously discussed. However, the prediction is performed again for a complete exemplification based on the sequence (E). The prediction below is considered for a period of 5 days. In the first instance, the sequence (E) is temporarily increased with five unknowns ("?????"), namely: E = "SRRSRSRRSRSRRS**?????**". In each stage, the unknowns in the sequence are gradually replaced (Figure 4.3).

Weather on Day 1

The initial state vector ($v^{(0)}$) is represented by the last known letter in the sequence, namely "**S**" ("SRRSRSRRSRSRRS**?????**"). Letter "**S**" denotes the current state of the system, namely the observation made today. Thus, the first prediction is made for day 1 based on the rules described previously above (Figure 4.3):

$$v^{(0)} \times P = v^{(1)}$$

$$[1 \quad 0] \begin{bmatrix} 0.2 & 0.8 \\ 0.625 & 0.375 \end{bmatrix} = [0.2 \quad 0.8]$$

Component x_1 represents the probability of a sunny day for tomorrow and component y_1 represents the probability of a rainy day for tomorrow.

$$v^{(1)} = [x_1 \quad y_1] = [0.2 \quad 0.8]$$

The vector components are evaluated. The assessment is made by determining which of the components contains a higher probability value. Component x_1 holds a value of 0.2 (sunny day), whereas component y_1 holds a value of 0.8 (rainy day):

$$\text{Day}(v^{(1)}) = \begin{cases} R, & x_1 < y_1 \\ S, & x_1 \geq y_1 \end{cases}$$

Component y_1 represents the rainy day and has a higher value than component x_1. Therefore, component y_1 indicates a rainy day as the most likely outcome on day 1. Following the example from the beginning, the first question mark ("**?**") from the sequence ("SRRSRSRRSRSRRS**?????**") can be replaced with an "**R**": E$_1$ = "SRRSRSRRSRSRRS**R????**".

Weather on Day 2

The prediction for day 2 is made as follows (Figure 4.3):

$$v^{(1)} \times P = v^{(2)}$$

$$[0.2 \quad 0.8] \begin{bmatrix} 0.2 & 0.8 \\ 0.625 & 0.375 \end{bmatrix} = [0.54 \quad 0.46]$$

$$v^{(2)} = [x_2 \quad y_2] = [0.54 \quad 0.46]$$

$$\text{Day}(v^{(2)}) = \begin{cases} R, & x_2 < y_2 \\ S, & x_2 \geq y_2 \end{cases}$$

Component x_2 contains a higher probability value than component y_2. Thus, component x_2 indicates a sunny day as the most likely outcome on day 2. Consequently, the second question mark ("**?**") from the sequence E$_1$ can be replaced with an "**S**": E$_2$ = "SRRSRSRRSRSRRS**RS???**".

Weather on Day 3

Accordingly, the prediction for day 3 is made in the same manner as above (Figure 4.3):

$$v^{(2)} \times P = v^{(3)}$$

$$[\,0.54 \quad 0.46\,] \begin{bmatrix} 0.2 & 0.8 \\ 0.625 & 0.375 \end{bmatrix} = [\,0.4 \quad 0.6\,]$$

$$v^{(3)} = [\,x_3 \quad y_3\,] = [\,0.4 \quad 0.6\,]$$

$$\mathrm{Day}(v^{(3)}) = \begin{cases} R, & x_3 < y_3 \\ S, & x_3 \geq y_3 \end{cases}$$

Since component $x_3 < y_3$, the third question mark ("?") from sequence E_2 can be replaced with an "R", namely: E_3 = "SRRSRSRRSRSRRS$\underline{R}$SR??".

Weather on Day 4

Next, the calculations are made for day 4 (Figure 4.3):

$$v^{(3)} \times P = v^{(4)}$$

$$[\,0.4 \quad 0.6\,] \begin{bmatrix} 0.2 & 0.8 \\ 0.625 & 0.375 \end{bmatrix} = [\,0.46 \quad 0.54\,]$$

$$v^{(4)} = [\,x_4 \quad y_4\,] = [\,0.46 \quad 0.54\,]$$

$$\mathrm{Day}(v^{(4)}) = \begin{cases} R, & x_4 < y_4 \\ S, & x_4 \geq y_4 \end{cases}$$

The component $x_4 < y_4$, therefore, the fourth question mark ("?") can be replaced from sequence E_3 with an "R": E_4 = "SRRSRSRRSRSRRS$\underline{S}$RSRR?".

Weather on Day 5

Next, the calculations are made for day 5 (Figure 4.3):

$$v^{(4)} \times P = v^{(5)}$$

$$[\,0.46 \quad 0.54\,] \begin{bmatrix} 0.2 & 0.8 \\ 0.625 & 0.375 \end{bmatrix} = [\,0.43 \quad 0.57\,]$$

$$v^{(5)} = [\,x_5 \quad y_5\,] = [\,0.43 \quad 0.57\,]$$

$$\mathrm{Day}(v^{(5)}) = \begin{cases} R, & x_5 < y_5 \\ S, & x_5 \geq y_5 \end{cases}$$

Since component $x_5 < y_5$, the last question mark ("?") can be replaced from sequence E_4 with an "R":

$$E_5 = \text{"SRRSRSRRSRSRRS}\underline{\text{S}}\text{RSRRR"}$$

In conclusion, at the beginning of this analysis, a hypothetical sequence of observations was considered for prediction, namely: E = "SRRSRSRRSRSRRSS".

A transition matrix has been created based on the transitions from each letter to each letter in the sequence. Next, the transition matrix has been used for a weather prediction on a period of 5 days. The implementation below uses the transition matrix and the initial state vector to make a weather prediction for a period of 5 days:

```
Private Sub Form_Load()
Dim P(1 To 2, 1 To 2) As Variant
Dim v(0 To 1) As Variant

chain = 5

P(1, 1) = 0.2
P(1, 2) = 0.625
P(2, 1) = 0.8
P(2, 2) = 0.375

v(0) = 1
v(1) = 0

For i = 1 To chain

    x = (v(0) * P(1, 1)) + (v(1) * P(1, 2))
    y = (v(0) * P(2, 1)) + (v(1) * P(2, 2))

    v(0) = x
    v(1) = y

    MsgBox v(0) & " | " & v(1)

Next i

End Sub
```

```
Output:
0.2 | 0.8
0.54 | 0.46
0.3955 | 0.6045
0.4569125 | 0.5430875
0.4308121875 | 0.5691878125
```

Supporting algorithm 4: Step-by-step prediction using a 2 × 2 transition matrix. A probability vector is repeatedly multiplied by a transition matrix. The vectors obtained from these repetitions show the probability of each outcome on a particular step.

An initial probability vector is multiplied by a transition matrix (Supporting algorithm 4). The resulting vector shows the probability of each outcome when the system takes one discrete step. The new vector is again multiplied by the transition matrix in order to show the probability of each outcome after two discrete steps. At each loop, the previous vector is multiplied by the matrix until the fifth prediction cycle is reached.

4.3 The Steady State of a Markov Chain

But how many days can be predicted in this way? Is there an end? Predictions for the weather on more distant days are increasingly inaccurate and tend toward a steady-state vector. The steady-state vector ($v^{(k)}$) of a transition matrix (P) is the unique probability vector that satisfies this equation:

$$v^{(k)} \times P = v^{(k)}$$

Different transition matrices allow for different number of iterations (or chains), but eventually the equilibrium is achieved. Thus, for a large number of days (k), the following expressions can be written:

$$v^{(0)} \times P = v^{(1)}$$
$$v^{(1)} \times P = v^{(2)}$$
$$v^{(2)} \times P = v^{(3)}$$
$$v^{(3)} \times P = v^{(4)}$$
$$v^{(4)} \times P = v^{(5)}$$
$$\vdots$$
$$v^{(k-1)} \times P = v^{(k)}$$
$$v^{(k)} \times P = v^{(k)}$$

The implementation below uses the transition matrix (P) and the initial state vector ($v^{(0)}$) to make a weather prediction for a period of 50 days. The output of this computer implementation can be observed in Table 4.1.

```
Private Sub Form_Load()
Dim P(1 To 2, 1 To 2) As Variant
Dim v(0 To 1) As Variant

chain = 50

P(1, 1) = 0.2
P(1, 2) = 0.625
P(2, 1) = 0.8
P(2, 2) = 0.375
```

```
v(0) = 1
v(1) = 0

For i = 1 To chain

    x = (v(0) * P(1, 1)) + (v(1) * P(1, 2))
    y = (v(0) * P(2, 1)) + (v(1) * P(2, 2))

    If v(0) = x And v(1) = y Then
        MsgBox "Steady state vector at day [" & i & "]!"
        Exit Sub
    Else
        MsgBox "Day[" & i & "], v=[" & x & " | " & y & "]"
    End If

    v(0) = x
    v(1) = y

Next i

End Sub
```

Supporting algorithm 5: Step-by-step prediction for 50 discrete steps. A probability vector is repeatedly multiplied by a 2 × 2 transition matrix. The vectors obtained from these repeated multiplications show the probability of each outcome on a particular step. Furthermore, at every cycle, the old vector is compared to the new vector in order to detect the first occurrence of equilibrium, namely the steady-state vector.

The results shown in Table 4.1 indicate that in this particular weather example, the prediction may be helpful for a period of 40 days. The equilibrium (steady-state vector) is achieved after 40 iterations:

$$v^{(0)} \neq v^{(1)} \neq \ldots \neq v^{(40)} = v^{(41)} = \ldots = v^{(50)}$$

Subsequent iterations through the transition matrix (P) do not change the values of the resulting vector (Table 4.1). A pertinent question would be if the steady-state vector could be found without the use of step-by-step iterations. As it has been shown previously, a state of equilibrium is reached if a vector multiplied by a transition matrix shows the same vector:

$$v \times P = v$$

Starting from the above observation, v can be subtracted from both sides of the equation and the following can be written:

$$v \times P - v = 0$$

Table 4.1 Step-by-step weather prediction for 50 days.

Day	State	$v = [x \quad y]$
[1]	R	[0.2 | 0.8]
[2]	S	[0.54 | 0.46]
[3]	R	[0.3955 | 0.6045]
[4]	R	[0.4569125 | 0.5430875]
[5]	R	[0.4308121875 | 0.5691878125]
[6]	R	[0.4419048203125 | 0.5580951796875]
[7]	R	[0.437190451367188 | 0.562809548632813]
[8]	R	[0.439194058168945 | 0.560805941831055]
[9]	R	[0.438342525278198 | 0.561657474721802]
[10]	R	[0.438704426756766 | 0.561295573243234]
[11]	R	[0.438550618628375 | 0.561449381371626]
[12]	R	[0.438615987082941 | 0.561384012917059]
[13]	R	[0.438588205489750 | 0.561411794510250]
[14]	R	[0.438600012666856 | 0.561399987333144]
[15]	R	[0.438594994616586 | 0.561405005383414]
[16]	R	[0.438597127287951 | 0.561402872712049]
[17]	R	[0.438596220902621 | 0.561403779097379]
[18]	R	[0.438596606116386 | 0.561403393883614]
[19]	R	[0.438596442400536 | 0.561403557599464]
[20]	R	[0.438596511979772 | 0.561403488020228]
[21]	R	[0.438596482408597 | 0.561403517591403]
[22]	R	[0.438596494976347 | 0.561403505023654]
[23]	R	[0.438596489635053 | 0.561403510364948]
[24]	R	[0.438596491905103 | 0.561403508094898]
[25]	R	[0.438596490940332 | 0.561403509059669]
[26]	R	[0.438596491350359 | 0.561403508649641]
[27]	R	[0.438596491176098 | 0.561403508823903]
[28]	R	[0.438596491250159 | 0.561403508749842]
[29]	R	[0.438596491218683 | 0.561403508781318]
[30]	R	[0.438596491232060 | 0.561403508767940]
[31]	R	[0.438596491226375 | 0.561403508773626]
[32]	R	[0.438596491228791 | 0.561403508771209]

Table 4.1 (*Continued*)

Day	State	$v = [x \quad y]$
[33]	R	[0.438596491227764 | 0.561403508772236]
[34]	R	[0.438596491228201 | 0.561403508771800]
[35]	R	[0.438596491228015 | 0.561403508771986]
[36]	R	[0.438596491228094 | 0.561403508771907]
[37]	R	[0.438596491228060 | 0.561403508771940]
[38]	R	[0.438596491228075 | 0.561403508771926]
[39]	R	[0.438596491228069 | 0.561403508771932]
[40]	R	**[0.438596491228071 | 0.56140350877193]**
[41]	R	[0.438596491228071 | 0.56140350877193]
[42]	R	[0.438596491228071 | 0.56140350877193]
[43]	R	[0.438596491228071 | 0.56140350877193]
[44]	R	[0.438596491228071 | 0.56140350877193]
[45]	R	[0.438596491228071 | 0.56140350877193]
[46]	R	[0.438596491228071 | 0.56140350877193]
[47]	R	[0.438596491228071 | 0.56140350877193]
[48]	R	[0.438596491228071 | 0.56140350877193]
[49]	R	[0.438596491228071 | 0.56140350877193]
[50]	R	[0.438596491228071 | 0.56140350877193]

Steady-state vector at [40]

Next, a factorization is further performed for v on the left side of the equation:

$$v(P - 1) = 0$$

The steady-state vector (v) is independent from the initial state vector and is unchanged when transformed by P. This means v is an eigenvector with eigenvalue 1, and it can be derived from P. In the next stage, an identity matrix is introduced. In matrix multiplication, an identity matrix (I) is equivalent to 1, therefore the previous equation becomes:

$$v(P - I) = 0$$

Up to this point, the components of v (the steady-state vector) consist of two unknowns, namely x and y:

$$v = [x \quad y]$$

However, the elements of the transition probability matrix (P) are known from the previous example:

$$P = \begin{bmatrix} 0.2 & 0.8 \\ 0.625 & 0.375 \end{bmatrix}$$

Also, the identity matrix (I) can be written as:

$$I = \begin{bmatrix} 1 & 0 \\ 0 & 1 \end{bmatrix}$$

Since x and y are the components of a probability vector, their sum is equal to 1 ($x + y = 1$). Therefore, v, P, and I are placed in context:

$$[x \quad y]\left(\begin{bmatrix} 0.2 & 0.8 \\ 0.625 & 0.375 \end{bmatrix} - \begin{bmatrix} 1 & 0 \\ 0 & 1 \end{bmatrix}\right) = 0$$

$$[x \quad y]\left(\begin{bmatrix} 0.2 - 1 & 0.8 - 0 \\ 0.625 - 0 & 0.375 - 1 \end{bmatrix}\right) = 0$$

$$[x \quad y]\begin{bmatrix} -0.8 & 0.8 \\ 0.625 & -0.625 \end{bmatrix} = 0$$

We multiply the vector with the matrix and obtain:

$$[x \quad y]\begin{bmatrix} -0.8 & 0.8 \\ 0.625 & -0.625 \end{bmatrix} = 0$$

$$-0.8x + 0.625y = 0$$

$$0.8x - 0.625y = 0$$

If the first equation ($-0.8x + 0.625y = 0$) is multiplied by -1, the second equation is obtained ($0.8x - 0.625y = 0$). Therefore, since the two equations are identical, only one of them is used further for determining the steady-state vector ($0.8x - 0.625y = 0$). As it was stated above, x and y are the components of a probability vector and their sum is equal to 1 ($x + y = 1$). Thus, a system of equations can be written as follows:

$$\begin{cases} x + y = 1 \\ 0.8x - 0.625y = 0 \end{cases}$$

$$x = \frac{25}{57} = 0.43859649122807$$

$$y = \frac{32}{57} = 0.56140350877193$$

$$v = [0.44 \quad 0.56]$$

In order to verify the steady-state vector, note that a multiplication of the vector (v) with the matrix (P) leads to the same result:

$$v \times P = v$$

$$[0.44 \quad 0.56]\begin{bmatrix} 0.2 & 0.8 \\ 0.625 & 0.375 \end{bmatrix} = [0.44 \quad 0.56]$$

The system of equations:

$$\begin{cases} x + y = 1 \\ 0.8x - 0.625y = 0 \end{cases}$$

Solve $x + y = 1$ for x. Add $-y$ to both sides:

$$x + y + (-y) = 1 + (-y)$$
$$x = -y + 1$$

Substitute $-y + 1$ for x in $0.8x - 0.625y = 0$:

$$0.8x - 0.625y = 0$$
$$0.8(-y + 1) - 0.625y = 0$$
$$-1.425y + 0.8 = 0$$

Simplify both sides of the equation. Add -0.8 to both sides:

$$-1.425y + 0.8 + (-0.8) = 0 + (-0.8)$$
$$-1.425y = -0.8$$

Divide both sides by -1.425:

$$\frac{-1.425y}{-1.425} = \frac{-0.8}{-1.425}$$
$$y = 0.561404$$

Substitute 0.561404 for y in $x = -y + 1$:

$$x = -y + 1$$
$$x = -0.561404 + 1$$
$$x = 0.438596$$

Thus, the vector components are:

$$x = 0.438596$$
$$y = 0.561404$$

Therefore, the steady-state vector (v) has been found without making step-by-step iterations. Note that the steady-state vector is independent of the initial state vector. The steady-state vector represents the probabilities of sunny (x component) and rainy (y component) weather and indicates in the long term that about 44% of days will be sunny and ~56% of days will be rainy. The convergence toward the steady-state vector represents the natural path of the machine, thus allowing a weather prediction from day to day (from state to state). Largely, the prediction is an assumption that the convergence path of this numerical system will overlap actual paths of events. In nature, an enormous number of variables are involved. Consequently, the prediction will overlap real events on short periods of time. Another, more direct approach to the calculation of the steady-state vector is shown below:

$$P = \begin{bmatrix} 0.2 & 0.8 \\ 0.625 & 0.375 \end{bmatrix} = \begin{bmatrix} \alpha & (1-\beta) \\ (1-\alpha) & \beta \end{bmatrix}$$

$\alpha = 0.2$

$\beta = 0.375$

$v = \begin{bmatrix} x & y \end{bmatrix}$

The first component of the vector is deduced by:

$$x = \frac{(1-\beta)}{(2-(\alpha+\beta))}$$

$$x = \frac{(1-0.375)}{(2-(0.2+0.375))}$$

$$x = 0.438596491228071$$

$$v = \begin{bmatrix} 0.44 & y \end{bmatrix}$$

The second component of the vector is deduced by:

$$y = \frac{(1-\alpha)}{(2-(\alpha+\beta))}$$

$$y = \frac{(1-0.2)}{(2-(0.2+0.375))}$$

$$y = 0.56140350877193$$

$$v = \begin{bmatrix} 0.44 & 0.56 \end{bmatrix}$$

Thus, again the equilibrium is satisfied and the above formulas have led directly to the point where the chain is showing a complete relaxation:

$$v \times P = v$$

$$\begin{bmatrix} 0.44 & 0.56 \end{bmatrix} \begin{bmatrix} 0.2 & 0.8 \\ 0.625 & 0.375 \end{bmatrix} = \begin{bmatrix} 0.44 & 0.56 \end{bmatrix}$$

The implementation below uses the transition matrix (P) to obtain the steady-state vector:

```
Private Sub Form_Load()
Dim P(1 To 2, 1 To 2) As Variant
Dim v(0 To 1) As Variant

P(1, 1) = 0.2
P(1, 2) = 0.625
P(2, 1) = 0.8
P(2, 2) = 0.375
```

```
a = P(1, 1)
b = P(2, 2)

x = (1 - b) / (2 - (a + b))
y = (1 - a) / (2 - (a + b))

v(0) = x
v(1) = y

MsgBox "Steady state vector v = [" & v(0) & " | " & v(1) & "]"

End Sub
```

```
Output:
Steady state vector v = [0.43859649122807 | 0.56140350877193]
```

Supporting algorithm 6: The computation of the steady-state vector. The above formulas are used for computing the *x* and *y* components of the steady-state vector. Note that iterations are not required.

Two methods for determining the steady-state vector have been evaluated. The first method uses a system of equations and the second method uses a specialized formula. Both methods resort to shortcuts that avoid step-by-step iterations and lead directly to the steady-state vector. Also, two algorithms have been used in order to make a prediction based exclusively on a sequence of observations (Supporting algorithm 3 and Supporting algorithm 4). However, these two implementations have been discussed in separate contexts. The implementation below integrates Supporting algorithm 3 and Supporting algorithm 4 in the same context based on all those discussed so far (Supporting algorithm 7).

```
Dim M(1 To 2, 1 To 2) As String

Private Sub Form_Load()
Dim v(0 To 1) As Variant

Call ExtractProb("SRRSRSRRSRSRRSS")

chain = 5

v(0) = 1
v(1) = 0
```

```
For i = 1 To chain

    x = (v(0) * M(1, 1)) + (v(1) * M(2, 1))
    y = (v(0) * M(1, 2)) + (v(1) * M(2, 2))

    v(0) = x
    v(1) = y

    MsgBox "Day (" & i & ")=[" & v(0) & " - " & v(1) & "]"

Next i
End Sub

Function ExtractProb(ByVal s As String)

Eb = "S"
Es = "R"

For i = 1 To 2
    For j = 1 To 2
      M(i, j) = 0
    Next j
Next i

TB = 0
TS = 0

For i = 2 To Len(s) - 1

        DI1 = Mid(s, i, 1)
        DI2 = Mid(s, i + 1, 1)

        If DI1 = Eb Then r = 1
        If DI1 = Es Then r = 2
        If DI2 = Eb Then c = 1
        If DI2 = Es Then c = 2

        M(r, c) = Val(M(r, c)) + 1

        If DI1 = Eb Then TB = TB + 1
        If DI1 = Es Then TS = TS + 1

Next i
```

```
For i = 1 To 2
    For j = 1 To 2
        If i = 1 Then M(i, j) = Val(M(i, j)) / TB
        If i = 2 Then M(i, j) = Val(M(i, j)) / TS
    Next j
Next i


End Function
```

```
Output:
Day (1)=[0.2 - 0.8]
Day (2)=[0.54 - 0.46]
Day (3)=[0.3955 - 0.6045]
Day (4)=[0.4569125 - 0.5430875]
Day (5)=[0.4308121875 - 0.5691878125]
```

Supporting algorithm 7: Step-by-step prediction using a sequence of observations made by a two-state Markov machine. First, a 2×2 matrix is used for counting all the combinations of pairs of letters ($D_{a \to b}$) in the sequence ($D_{a \to b}$ is represented by joining two string variables, namely DI1 and DI2). In parallel, the first letter (N_a) of each pair is counted inside the sequence (N_a is represented by variable DI1). Second, the transition probabilities are computed. The values from each element of the matrix are divided by the corresponding N_a. In the final phase, a probability vector is repeatedly multiplied by the new transition matrix. The vectors obtained from these repetitions show the probability of each outcome on a particular step.

The above algorithm brings forward a method that starts with the sequence of observations and ends with the prediction (Supporting algorithm 7). First, a transition matrix has been created based on a sequence of observations, namely E = "SRRSRSRRSRSRRSS". Second, the transition matrix and the initial state vector are used for a 5 days weather prediction.

4.4   The Long-Run Distribution of a Markov Chain

A Markov chain reveals the behavior of a random process, hence the power of prediction for upcoming events. Nevertheless, as the chain relaxes to a stable distribution, the predictive power of the method decreases. The convergence (relaxation) toward the steady-state vector represents the natural path of the machine to equilibrium. Here, some examples of chains are given in order to illustrate this relaxation. Up to this point, the components have been displayed as numeric values. However, vector components can also be visualized in a different manner. The values of the components may be plotted on a graph each time a discrete step is made. This approach shows an overview for the long-run path (convergence toward the steady-state vector) of the system (Figure 4.4).

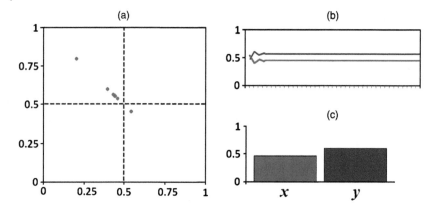

Figure 4.4 The long-run distribution of a two-state Markov chain. (a) The vertical axis represents the probability value of the *x* component and the horizontal axis represents the probability value of the *y* component. Each circle has the *x* and *y* coordinates dictated by the two components of the probability vector. (b) The convergence path toward the steady-state vector. The vertical axis represents the probability value for both vector components and the horizontal axis represents the long-run behavior of the vector components over 40 steps (*k* iterations). The red line represents the *x* component and the blue line represents the *y* component, (c) the red bar represents the *x* component and the blue bar represents the *y* component of the steady-state vector (equilibrium). The vertical axis represents the probability values.

In our previous example, the weather prediction has been made step by step for a period of 50 days (Supporting algorithm 5). However, the prediction showed that after day 40, the system reaches the steady-state vector (Table 4.1). Interestingly, when the components shown in Table 4.1 are plotted on a graph, an overview of the system behavior can be observed. Namely, the values of the vector components fluctuate a few steps before entering a smooth path toward the steady-state vector (Figures 4.4a and 4.4b). Nevertheless, different transition matrices allow for a different number of iterations (or chains), but eventually the equilibrium is achieved if all states are reachable (communicate with each other). Consequently, this convergence path can be slightly different for each case. Thus, to understand the convergence of a Markov chain toward the steady-state vector, several cases are considered (Table 4.2). In the examples below, two of the cases (Tables 4.2a and 4.2b) include a double stochastic matrix and the other three cases (Tables 4.2c–e) include three right stochastic matrices.

4.4.1 Case A

A double stochastic matrix is used in a Markov chain of 20 steps (Table 4.2a). The transition matrix is composed of two probability vectors. The probability

Table 4.2 Analysis of different settings. In the long-run distribution column, the vertical axis represents the probability and the horizontal axis represents the long-run behavior of the vector components. The red line represents the *x* component and the blue line represents the *y* component of the probability vector over several steps (iterations). In the steady-state vector column, the red bar represents the *x* component and the blue bar represents the *y* component of the steady-state vector (equilibrium). The vertical axis represents the probability values.

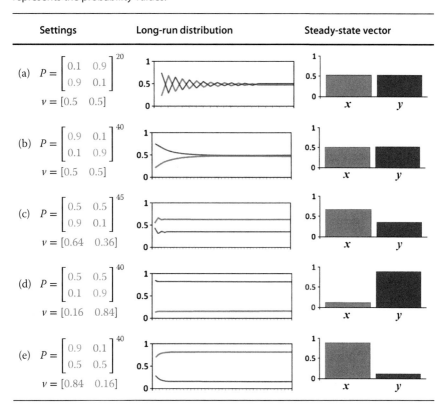

Settings	Long-run distribution	Steady-state vector
(a) $P = \begin{bmatrix} 0.1 & 0.9 \\ 0.9 & 0.1 \end{bmatrix}^{20}$ $v = [0.5 \quad 0.5]$		
(b) $P = \begin{bmatrix} 0.9 & 0.1 \\ 0.1 & 0.9 \end{bmatrix}^{40}$ $v = [0.5 \quad 0.5]$		
(c) $P = \begin{bmatrix} 0.5 & 0.5 \\ 0.9 & 0.1 \end{bmatrix}^{45}$ $v = [0.64 \quad 0.36]$		
(d) $P = \begin{bmatrix} 0.5 & 0.5 \\ 0.1 & 0.9 \end{bmatrix}^{40}$ $v = [0.16 \quad 0.84]$		
(e) $P = \begin{bmatrix} 0.9 & 0.1 \\ 0.5 & 0.5 \end{bmatrix}^{40}$ $v = [0.84 \quad 0.16]$		

vector from the first row of the transition matrix has the following two components: $P[S|S] = 0.1$ and $P[S|R] = 0.9$. Note that a transition probability is written as $P[\text{from}|\text{to}]$. The probability vector from the second row of the transition matrix has the following two components: $P[R|S] = 0.9$ and $P[R|R] = 0.1$.

$$[0.1 \quad 0.9]\begin{bmatrix} 0.1 & 0.9 \\ 0.9 & 0.1 \end{bmatrix}^{20} = [0.5 \quad 0.5]$$

The steady-state vector indicates that in the long run, 50% of the time will be sunny (in red) and 50% of the time will be rainy (in blue). However, the major observation is that at every step, the machine is switching from one state to the other state (Table 4.2a).

4.4.2 Case B

A double stochastic matrix is used in a Markov chain of 40 steps (Table 4.2b). The transition matrix is composed of two probability vectors. The probability vector from the first row of the transition matrix has the following two components: $P[S|S] = 0.9$ and $P[S|R] = 0.1$. The probability vector from the second row of the transition matrix has the following two components: $P[R|S] = 0.1$ and $P[R|R] = 0.9$.

$$[0.1 \quad 0.9] \begin{bmatrix} 0.9 & 0.1 \\ 0.1 & 0.9 \end{bmatrix}^{40} = [0.5 \quad 0.5]$$

Thus, the values of the vector components are switched compared to the first case. Also, the steady-state vector indicates that in the long run, 50% of the time will be sunny (in red) and 50% of the time will be rainy (in blue). Interestingly, the major observation in this case is that the machine slowly converges to the steady-state vector without changing the state (Table 4.2b).

4.4.3 Case C

A right stochastic matrix is used in a Markov chain of 45 steps (Table 4.2c). The transition matrix is composed of two probability vectors. The probability vector from the first row of the transition matrix has the following two components: $P[S|S] = 0.5$ and $P[S|R] = 0.5$. The probability vector from the second row of the transition matrix has the following two components: $P[R|S] = 0.9$ and $P[R|R] = 0.1$.

$$[0.1 \quad 0.9] \begin{bmatrix} 0.5 & 0.5 \\ 0.9 & 0.1 \end{bmatrix}^{45} = [0.64 \quad 0.36]$$

In this case, the steady-state vector indicates that in the long run, 64% of the time will be sunny (in red) and 36% of the time will be rainy (in blue). Notice that the values of the vector components fluctuate a few steps before entering a smooth path toward the steady state (Table 4.2c). This case shows a similar pattern to the example used so far along this chapter (Figure 4.4).

4.4.4 Case D

A right stochastic matrix is used in a Markov chain of 40 steps (Table 4.2d). The transition matrix is composed of two probability vectors. The probability vector from the first row of the transition matrix has the following two components: $P[S|S] = 0.5$ and $P[S|R] = 0.5$. The probability vector from the second row of the transition matrix has the following two components: $P[R|S] = 0.1$ and $P[R|R] = 0.9$.

$$[0.1 \quad 0.9]\begin{bmatrix} 0.5 & 0.5 \\ 0.1 & 0.9 \end{bmatrix}^{40} = [0.16 \quad 0.84]$$

In this case, the steady-state vector indicates that in the long run, 16% of the time will be sunny (in red) and 84% of the time will be rainy (in blue). The only change from the previous case is that the values of the vector components from the second row are switched. Here, the observation is that the machine behaves radically different by making a smooth entry into the steady state (Table 4.2d).

4.4.5 Case E

A right stochastic matrix is used in a Markov chain of 40 steps (Table 4.2e). The transition matrix is composed of two probability vectors. The probability vector from the first row of the transition matrix has the following two components: $P[S|S] = 0.9$ and $P[S|R] = 0.1$. The probability vector from the second row of the transition matrix has the following two components: $P[R|S] = 0.5$ and $P[R|R] = 0.5$.

$$[0.1 \quad 0.9]\begin{bmatrix} 0.9 & 0.1 \\ 0.5 & 0.5 \end{bmatrix}^{40} = [0.84 \quad 0.16]$$

In this case, the row vectors from the previous case have been switched. This time the steady-state vector indicates that in the long run, 84% of the time will be sunny (in red) and 16% of the time will be rainy (in blue). Compared to the previous example (case C), in this particular case, the row vectors are switched and the machine strengthens the likelihood of being in the sunny state. For more experiments related to convergence, please see the software implementation of the Markov chain simulator (Additional materials online).

5

Predictions Using n-State Markov Chains

5.1 Introduction

During the past decade, there has been an explosion of raw data from all areas of science [26, 27]. The new challenge for the information technology field consists in adapting and expanding well-known computation methods depending on the type of data [28–31]. Vast amounts of information in a variety of fields (such as biology, chemistry, medicine, finance, or marketing) led to the development of new computational tools [32, 33]. Each tool is most often specific to a well-defined problem. In the same way, new implementations of Markov chains in different analysis tools are unique (particularly those used in research). The parameters of such a system can vary from the number of states up to the relationship between states. For instance, the states of a Markov chain reflect the number of types of events present in a sequence of observations. This chapter describes some examples with more than two states. The first two examples include a three-state Markov chain and a four-state Markov chain, after which a gradual generalization is made for an arbitrary number of states (n states). The supporting theory is accompanied by an algorithm implementation for each example, which allows a prediction based on a sequence of observations.

5.2 Predictions by Using the Three-State Markov Chain

The two-state Markov chain has been the main focus for understanding the stochastic processes. Earlier in this chapter, some examples of black and white jars (each filled with black and white balls) have been gradually transformed into a Markov diagram (Figure 1.1). These gradual transformations can be extended for an example which contains three states (Figure 5.1a). In this new case, each state is represented by jars of different colors, namely yellow, blue, and gray. Also, each jar contains balls of three colors, namely yellow, blue, and gray. A series of draws are made by respecting the chain rule. Thus, the

Markov Chains: From Theory to Implementation and Experimentation, First Edition. Paul A. Gagniuc.
© 2017 John Wiley & Sons, Inc. Published 2017 by John Wiley & Sons, Inc.
Companion website: www.wiley.com/go/gagniuc/markovchains

(a)   (b)

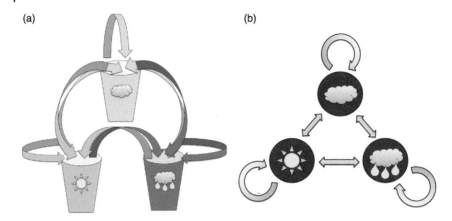

Figure 5.1 A three-state Markov chain. (a) Shows how the three-jar system is equivalent to the (b) Markov diagram representing the weather, by changing the angle of viewing of the jars from the side view to the top view.

proportion of balls in each of the three jars can be determined by observing the color of individual balls on a large number of draws. Since dependent variables are involved, the proportion of balls in each jar is determined by analyzing the transitions from one jar to another jar, or from a jar to the same jar. Once the proportion of balls is determined for each of the three jars, the behavior of this system can be predicted for future draws. However, "draws" cannot be performed in the case of the weather system as in the case of the jars system. Nevertheless, the weather events can be observed in order to predict the behavior of the weather system. Thus, the prediction can be slightly refined by considering a new observation. In addition to "sunny" and "rainy", the "cloudy" observation can also be added. Therefore, the new system accommodates three states, namely "S", "R", and "C" The manner in which this three-state system can be analyzed is no different from the one with two states. The Markov process captures the global rules of a natural system, disregarding the multitude of refined factors (variables) that led to those events. Therefore, the Markov process is the same whether the observations consist of "sunny", "rainy" and "cloudy", or of yellow, blue, and gray balls. Consider a new sequence (E) of observations which includes the third state. Suppose that the sequence of observations (E) contains 14 observations:

E = "SRCCRRSSCSRCSR"

In order to extract the transition probabilities, the number of transitions from state to state has to be counted. Three states (sunny, rainy, and cloudy) involve a total of nine transitions. From the state diagram, a 3×3 transition matrix can be formed:

$$P = \begin{pmatrix} P[S|S] & P[S|R] & P[S|C] \\ P[R|S] & P[R|R] & P[R|C] \\ P[C|S] & P[C|R] & P[C|C] \end{pmatrix}$$

The transition probabilities ($P[\text{from}|\text{to}]$) are calculated in a similar manner to that used in a system of two states, namely:

$$T_{a \to b} = \frac{\text{Count}(D_{a \to b})}{\text{Count}(N_a)}$$

where $T_{a \to b}$ is the transition probability from a state (a) to a state (b), $D_{a \to b}$ represents the number of transitions from a to b, and N_a represents the number of times a appears in the sequence. For instance, if individual states (N_a) and the transitions between states ($D_{a \to b}$) are counted inside the sequence (E), the following can be written:

$$P[S|S] = T_{S \to S} = \frac{\text{Count}(D_{S \to S})}{\text{Count}(N_S)} = \frac{1}{4} = 0.25$$

$$P[S|R] = T_{S \to R} = \frac{\text{Count}(D_{S \to R})}{\text{Count}(N_R)} = \frac{2}{4} = 0.5$$

$$P[S|C] = T_{S \to C} = \frac{\text{Count}(D_{S \to C})}{\text{Count}(N_C)} = \frac{1}{4} = 0.25$$

$$P[R|S] = T_{R \to S} = \frac{\text{Count}(D_{R \to S})}{\text{Count}(N_R)} = \frac{1}{4} = 0.25$$

$$\vdots$$

$$P[C|R] = T_{C \to R} = \frac{\text{Count}(D_{C \to R})}{\text{Count}(N_C)} = \frac{1}{4} = 0.25$$

$$P[C|C] = T_{C \to C} = \frac{\text{Count}(D_{C \to C})}{\text{Count}(N_C)} = \frac{1}{4} = 0.25$$

Note that $P[a|b]$ and $T_{a \to b}$ have the exact same meaning, only the notation is different. A detailed counting of the states (N_a) and the transitions between states ($D_{a \to b}$) is shown in a graphical manner in Table 5.1. Here, the underline shows the letters (N_a) that have been counted, and the orange color shows the transitions between letters ($D_{a \to b}$) that have been counted. Since the first and the last letter in the sequence are not counted, they are represented with a strikethrough line (where appropriate—in our case the sequence starts with "S" and ends with "R", therefore, the first letter is not counted for $T_{S \to S}$, $T_{S \to R}$, $T_{S \to C}$, and the last letter is not counted for $T_{R \to S}$, $T_{R \to R}$, $T_{R \to C}$). The only transition which is not counted is between the first and the second state (letter) and is represented with a strikethrough line through the first two letters (where

Table 5.1 From observations to transition probabilities.

The observed sequence	Markov diagram

E = "SRCCRRSSCSRCSR"

$P[S|S] = $SRCCRR$\underline{SS}C\underline{S}RC\underline{S}$R = 1/4 = 0.25

$P[S|R] = \underline{S}$RCCRRSSC$\underline{SR}$C$\underline{SR}$ = 2/4 = 0.5

$P[S|C] = $SRCCRRS$\underline{SC}SRC\underline{S}$R = 1/4 = 0.25

$P[R|S] = S\underline{RC}CR\underline{RS}SCS\underline{RC}S\underline{R}$ = 1/4 = 0.25

$P[R|R] = $SRCC$\underline{RR}$SSCSRCSR = 1/4 = 0.25

$P[R|C] = S\underline{RC}$CRRSSCS$\underline{RC}S\underline{R}$ = 2/4 = 0.5

$P[C|S] = SR\underline{CC}$RRSS$\underline{C}SR\underline{C}$SR = 2/4 = 0.5

$P[C|R] = SR\underline{CC}$RRSS$\underline{C}SR\underline{C}$SR = 1/4 = 0.25

$P[C|C] = SR\underline{CC}$RRSSCSRCSR = 1/4 = 0.25

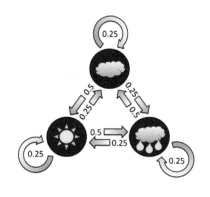

appropriate—in our case, the sequence starts with "SR"; therefore, the first transition is not counted for $T_{S \rightarrow R}$). Next, the probability values are inserted in the matrix (P):

$$P = \begin{pmatrix} P[S|S] & P[S|R] & P[S|C] \\ P[R|S] & P[R|R] & P[R|C] \\ P[C|S] & P[C|R] & P[C|C] \end{pmatrix} = \begin{pmatrix} 1/4 & 2/4 & 1/4 \\ 1/4 & 1/4 & 2/4 \\ 2/4 & 1/4 & 1/4 \end{pmatrix} = \begin{pmatrix} 0.25 & 0.5 & 0.25 \\ 0.25 & 0.25 & 0.5 \\ 0.5 & 0.25 & 0.25 \end{pmatrix}$$

Note that a transition probability is written as P[from|to]. Based on the sequence of observations (E), the transition matrix (P) was built and the prediction can be made. Let us consider a weather prediction for the next three days in a row. Thus, sequence (E) is temporarily increased with three unknowns ("???"):

$$E = \text{"SRCCRRSSCSRC}\underline{S}\underline{R}???\text{"}$$

The initial state vector ($v^{(0)}$) is built from the current observation. Three states are represented by a probability vector ($v^{(0)}$) with three components, namely:

$$v^{(0)} = [x_0 \quad y_0 \quad z_0]$$

where the x_0 component is the probability of being in the state "sunny", the y_0 component is the probability of being in the state "rainy", and the z_0 component is the probability of being in the state "cloudy":

$$v^{(0)} = [\text{sunny} \quad \text{rainy} \quad \text{cloudy}]$$

Since the last state (taken as today) in the sequence (E) is rainy (**R**), the probability of rainy is 1. The sum of the component values of a probability vector make unity; therefore, the other two components have zero probability:

$$v^{(0)} = [0 \quad 1 \quad 0]$$

A new vector ($v^{(1)}$) is obtained by using the transition matrix (P) and the probability vector ($v^{(0)}$) from above. The new vector ($v^{(1)}$) indicates the likelihood of being sunny, rainy, or cloudy tomorrow:

$$v^{(1)} = [x_1 \quad y_1 \quad z_1] = [? \quad ? \quad ?]$$

In order to calculate this new vector ($v^{(1)}$), the initial state vector ($v^{(0)}$) is multiplied by the transition matrix (P):

$$v^{(0)} \times P = v^{(1)}$$

To start the prediction of the events for tomorrow, the coordinates of the transition matrix are used as (m_{ij}):

$$[x_0 \quad y_0 \quad z_0] \begin{pmatrix} m_{11} & m_{12} & m_{13} \\ m_{21} & m_{22} & m_{23} \\ m_{31} & m_{32} & m_{33} \end{pmatrix} = [x_1 \quad y_1 \quad z_1]$$

The colors are meant to help us calculate the resulting vector. Note that in the case of the initial probability vector ($v^{(0)}$), the orange color represents the first component and is associated with the "sunny" state. The blue color represents the second component of this vector and is associated with the "rainy" state, and the gray color is the third component of this vector and is associated with the "cloudy" state. The colors of each component from the resulting vector ($v^{(1)}$) have the same meaning. The transition matrix is made also of three row probability vectors representing the state transitions. In this case, the orange color is associated with the first row probability vector that represents the transitions from "sunny" to another state (i.e., from "S" to "S", from "S" to "R", or from "S" to "C"). The blue color is associated with the second row probability vector and represents the transitions from "rainy" to another state. And last, the gray color is associated with the third row probability vector which represents the transitions from "cloudy" to another state (Table 5.2).

Weather on Day 1
The initial probability vector ($v^{(0)}$) and the transition matrix (P) can be placed in context:

$$[0 \quad 1 \quad 0] \begin{pmatrix} 0.25 & 0.5 & 0.25 \\ 0.25 & 0.25 & 0.5 \\ 0.5 & 0.25 & 0.25 \end{pmatrix} = [? \quad ? \quad ?]$$

Table 5.2 Step-by-step calculation of the vector components for day 1.

x_1 component	y_1 component	z_1 component
$x_1 = (x_0 \times m_{11})$ $+ (y_0 \times m_{21}) + (z_0 \times m_{31})$	$y_1 = (x_0 \times m_{12})$ $+ (y_0 \times m_{22}) + (z_0 \times m_{32})$	$z_1 = (x_0 \times m_{13})$ $+ (y_0 \times m_{23}) + (z_0 \times m_{33})$
$x_1 = (0 \times 0.25)$ $+ (1 \times 0.25) + (0 \times 0.5)$	$y_1 = (0 \times 0.5)$ $+ (1 \times 0.25) + (0 \times 0.25)$	$z_1 = (0 \times 0.25)$ $+ (1 \times 0.5) + (0 \times 0.25)$
$x_1 = (0) + (0.25) + (0)$	$y_1 = (0) + (0.25) + (0)$	$z_1 = (0) + (0.5) + (0)$
$x_1 = 0.25$	$y_1 = 0.25$	$z_1 = 0.5$

Thus, the likelihood of a sunny day tomorrow is represented by component x_1. The likelihood of a rainy day tomorrow is represented by the second component (y_1); and the likelihood of a cloudy day tomorrow is represented by the third component, namely z_1. Step-by-step calculation of these components is shown in Table 5.2. The components of the new probability vector ($v^{(1)}$) have been found:

$$v^{(1)} = [\,0.25 \quad 0.25 \quad 0.5\,]$$

In the next phase, the vector components are evaluated. The assessment is made by determining which of the components has a higher value. Component $x_1 = 0.25$ (sunny day), component $y_1 = 0.25$ (rainy day), and component $z_1 = 0.5$ (cloudy day). Component z_1 represents a cloudy day and has a higher value than component x_1 or y_1. Therefore, component z_1 indicates a cloudy day as the most likely outcome on day 1. The first question mark ("?") from the sequence "SRCCRRSSCSRCSR???" can be replaced with the letter "C": E = "SRCCRRSSCSRCSR<u>C</u>??"

Weather on Day 2

The prediction can be made for day 2 by following the same steps:

$$v^{(1)} \times P = v^{(2)}$$

$$[\,0.25 \quad 0.25 \quad 0.5\,] \begin{pmatrix} 0.25 & 0.5 & 0.25 \\ 0.25 & 0.25 & 0.5 \\ 0.5 & 0.25 & 0.25 \end{pmatrix} = [\,? \quad ? \quad ?\,]$$

Thus, the likelihood of being sunny the day after tomorrow is represented by the x_2 component, the likelihood of being rainy the day after tomorrow is represented by the y_2 component, and the likelihood of being cloudy the day after tomorrow is represented by the third component, namely z_2. Step-by-step

Table 5.3 Step-by-step calculation of the vector components for day 2.

x_2 component	y_2 component	z_2 component
$x_2 = (x_1 \times m_{11}) + (y_1 \times m_{21})$ $+ (z_1 \times m_{31})$	$y_2 = (x_1 \times m_{12}) + (y_1 \times m_{22})$ $+ (z_1 \times m_{32})$	$z_2 = (x_1 \times m_{13}) + (y_1 \times m_{23})$ $+ (z_1 \times m_{33})$
$x_2 = (0.25 \times 0.25) + (0.25 \times$ $0.25) + (0.5 \times 0.5)$	$y_2 = (0.25 \times 0.5) + (0.25 \times 0.25)$ $+ (0.5 \times 0.25)$	$z_2 = (0.25 \times 0.25) + (0.25 \times 0.5)$ $+ (0.5 \times 0.25)$
$x_2 = (0.0625) + (0.0625) + (0.25)$	$y_2 = (0.125) + (0.0625) + (0.125)$	$z_2 = (0.0625) + (0.125) + (0.125)$
$x_2 = 0.375$	$y_2 = 0.3125$	$z_2 = 0.3125$

calculation of these components is shown in Table 5.3. The components of the new probability vector ($v^{(2)}$) have been found:

$$v^{(2)} = [\,0.375 \quad 0.3125 \quad 0.3125\,]$$

The vector components are evaluated. The assessment is made by determining which of the components has a higher value. Component $x_1 = 0.375$, component $y_1 = 0.3125$, and component $z_1 = 0.3125$. Therefore, component x_1 has a higher value than component y_1 or component z_1. Thus, day 2 is more likely to be a sunny day. The second question mark ("**?**") can be replaced from the sequence "SRCCRRSSCSRCS<u>R</u>C**??**" with an "**S**": E="SRCCRRSSCSRCS<u>R</u>C**S**?"

Weather on Day 3

The same rules apply for day 3. Accordingly, a vector ($v^{(3)}$) indicates the likelihood of being sunny, rainy, or cloudy days in advance. The new vector ($v^{(3)}$) is obtained by using the values of the transition matrix (P) and the most recent probability vector ($v^{(2)}$) from above:

$$v^{(2)} \times P = v^{(3)}$$

The known values are placed in the new context:

$$[\,0.375 \quad 0.3125 \quad 0.3125\,] \begin{pmatrix} 0.25 & 0.5 & 0.25 \\ 0.25 & 0.25 & 0.5 \\ 0.5 & 0.25 & 0.25 \end{pmatrix} = [\,? \quad ? \quad ?\,]$$

Again, the likelihood of a sunny weather three days from now is represented by the x_3 component, the likelihood of a rainy weather three days from now is represented by the y_2 component, and the likelihood of a cloudy weather three days from now is represented by the z_2 component (Table 5.4). The components of the new probability vector ($v^{(3)}$) have been found:

$$v^{(3)} = [\,0.328125 \quad 0.34375 \quad 0.328125\,]$$

Table 5.4 Step-by-step calculation of the vector components for day 3.

x_3 component	y_3 component	z_3 component
$x_3 = (x_1 \times m_{11}) + (y_1 \times m_{21})$ $+ (z_1 \times m_{31})$	$y_3 = (x_1 \times m_{12}) + (y_1 \times m_{22})$ $+ (z_1 \times m_{32})$	$z_3 = (x_1 \times m_{13}) + (y_1 \times m_{23})$ $+ (z_1 \times m_{33})$
$x_3 = (.375 \times .25) + (.3125 \times .25)$ $+ (.3125 \times .5)$	$y_3 = (.375 \times .5) + (.3125 \times .25)$ $+ (.3125 \times .25)$	$z_3 = (.375 \times .25) + (.3125 \times .5)$ $+ (.3125 \times .25)$
$x_3 = (0.09375) + (0.078125)$ $+ (0.15625)$	$y_3 = (0.1875) + (0.078125)$ $+ (0.078125)$	$z_3 = (0.09375) + (0.15625)$ $+ (0.078125)$
$x_3 = 0.328125$	$y_3 = 0.34375$	$z_3 = 0.328125$

The vector components are evaluated. The evaluation is made by determining which of the components has a higher value. Component $x_1 = 0.328125$, $y_1 = 0.34375$, and component $z_1 = 0.328125$. Therefore, y_3 has a higher value than x_3 or z_3. It seems that day 3 is more likely to be a rainy day. Following the example from the beginning, the third question mark ("?") can be replaced from the sequence "SRCCRRSSCSRCS$\underline{R}$CS?" with an "R":

$$E = \text{"SRCCRRSSCS}\underline{R}\text{CSR"}$$

The sum of probabilities in a given situation must be equal to 1. Note that probability vectors with two components (x and y) have been used in all the previous cases. However, since in this case there are three states involved, the probability vector requires three components (x, y, z). In the case of probability vectors with two components, the value of one of the components dictated the value of the other component. This complementary was easily deducted by $1 - p$. Here, the values of two components must be known in order to find the value of a third component, for instance:

$$P[x] = 1 - P[y] + P[z]$$
$$P[y] = 1 - P[x] + P[z]$$
$$P[z] = 1 - P[x] + P[y]$$

Consequently, for probability vectors with a higher number (n) of components, the values of any $n - 1$ components must be known in order to deduce the value of any one component. The computer code below summarizes those discussed so far on the three-state Markov chain issue. First, the implementation computes the transition probabilities between letters (past observations) of "SRCCRRSSCSRCSR" and generates a 3×3 right stochastic matrix. In the second phase, the stochastic matrix is used in a weather prediction for a period of 5 days:

```
Dim M(1 To 3, 1 To 3) As String

Private Sub Form_Load()
Dim v(0 To 2) As Variant

Call ExtractProb("SRCCRRSSCSRCSR")

chain = 5
v(0) = 0
v(1) = 1
v(2) = 0

For i = 1 To chain

    x = (v(0) * M(1, 1)) + (v(1) * M(2, 1)) + (v(2) * M(3, 1))
    y = (v(0) * M(1, 2)) + (v(1) * M(2, 2)) + (v(2) * M(3, 2))
    z = (v(0) * M(1, 3)) + (v(1) * M(2, 3)) + (v(2) * M(3, 3))

    v(0) = x
    v(1) = y
    v(2) = z

    MsgBox "Day (" & i & ")=[" & v(0) & " | " & v(1) & " | "
& v(2) & "]"

Next i
End Sub

Function ExtractProb(ByVal s As String)

Eb = "S"
Es = "R"
Ec = "C"

For i = 1 To 3
    For j = 1 To 3
      M(i, j) = 0
    Next j
Next i

TB = 0
TS = 0
TC = 0
```

```
For i = 2 To Len(s) - 1

        DI1 = Mid(s, i, 1)
        DI2 = Mid(s, i + 1, 1)

        If DI1 = Eb Then r = 1
        If DI1 = Es Then r = 2
        If DI1 = Ec Then r = 3
        If DI2 = Eb Then c = 1
        If DI2 = Es Then c = 2
        If DI2 = Ec Then c = 3

        M(r, c) = Val(M(r, c)) + 1

        If DI1 = Eb Then TB = TB + 1
        If DI1 = Es Then TS = TS + 1
        If DI1 = Ec Then TC = TC + 1

Next i

For i = 1 To 3
    For j = 1 To 3
        If i = 1 Then M(i, j) = Val(M(i, j)) / TB
        If i = 2 Then M(i, j) = Val(M(i, j)) / TS
        If i = 3 Then M(i, j) = Val(M(i, j)) / TC
    Next j
Next i

End Function
```

```
Output:
Day (1)=[0.25 | 0.25 | 0.5]
Day (2)=[0.375 | 0.3125 | 0.3125]
Day (3)=[0.328125 | 0.34375 | 0.328125]
Day (4)=[0.33203125 | 0.33203125 | 0.3359375]
Day (5)=[0.333984375 | 0.3330078125 | 0.3330078125]
```

Supporting algorithm 8: Step-by-step prediction using a sequence of observations made by a three-state Markov machine. First, a 3×3 matrix is used for counting all the combinations of pairs of letters ($D_{a \to b}$) in the sequence ($D_{a \to b}$ is represented by joining two string variables, namely DI1 and DI2). In parallel, the first letter (N_a) of each pair is counted inside the sequence (N_a is represented by variable DI1). Second, the transition probabilities are computed. The values from each element of the matrix are divided by their corresponding N_a. In the final phase, a probability vector is repeatedly multiplied by the new transition matrix. The vectors obtained from these repetitions show the probability of each outcome on a particular step.

5.3 Predictions by Using the Four-State Markov Chain

All the examples have been based on the weather and the extractions of balls from jars. It seems that the weather (or the extraction of balls from jars) prediction may lead to a better understanding and a correct intuition regarding the stochastic process. All these steps involved in the weather prediction are particularly important for understanding other predictions made, for instance, in the field of genetics or bioinformatics [34]. The weather is dependent on a series of factors, from pressure, temperature, humidity, to the landscape. DNA sequences on the other hand (like the weather phenomena), are also dependent on many factors that dictate the order of nucleotides in the sequence over the evolution timeline [35, 36]. Therefore, DNA, RNA, or protein sequences enter in this category of dependent variables and may be analyzed in the same way as the weather [37, 38]. For instance, DNA is composed of four types of nucleotides, Adenine (A), Thymine (T), Cytosine (C), and Guanine (G). Therefore, there may be four types of observations in a sequence, namely "A", "T", "C", or "G". Four types of observations involve four states (Figure 5.2). Thus, for illustration, a DNA sequence can be regarded as a discrete-time Markov chain (DTMC). In a DTMC, the system evolves through discrete time steps and the changes to the system can only occur at one of those time intervals. Consider a new sequence (E) of observations which includes four states. Suppose that the sequence of observations (E) contains 31 observations as follows:

E = "TACTTCGATTTAAGCGCGGCGGCCTATATTA"

Figure 5.2 Markov diagram representing the DNA sequence.

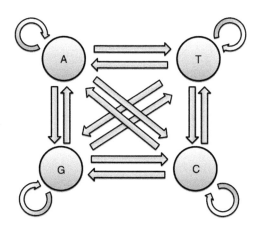

From the state diagram, a 4×4 transition probability matrix can be formed:

$$P = \begin{pmatrix} P[A|A] & P[A|T] & P[A|G] & P[A|C] \\ P[T|A] & P[T|T] & P[T|G] & P[T|C] \\ P[G|A] & P[G|T] & P[G|G] & P[G|C] \\ P[C|A] & P[C|T] & P[C|G] & P[C|C] \end{pmatrix}$$

Calculation of transition probabilities ($P[\text{from}|\text{to}]$) is made in a similar manner to that used in two states or three states:

$$T_{a \to b} = \frac{\text{Count}(D_{a \to b})}{\text{Count}(N_a)}$$

where a is the letter from which the transition is made, and b is the letter to which the transition is made. A detailed counting of the states (N_a) and the transitions between states ($D_{a \to b}$) are shown in a graphical manner in Table 5.5. Note again that $P[a|b]$ and $T_{a \to b}$ have the same meaning, only the notation is different. In Table 5.5, the underline shows the letters (N_a) that have been counted, and the orange color shows the transitions between letters ($D_{a \to b}$) that have been counted. Since the first and the last letter in the sequence are not counted, they are represented with a strikethrough line (where appropriate). In our case, the sequence starts with "T" and ends with "A"; therefore, the first letter is not counted for $T_{T \to A}$, $T_{T \to T}$, $T_{T \to C}$, $T_{T \to G}$, and the last letter is not counted for $T_{A \to A}$, $T_{A \to T}$, $T_{A \to C}$, $T_{A \to G}$. Also, the only transition which is not counted is between the first and the second state (letter) and is represented with a strikethrough line through the first two letters (where appropriate). Here, the sequence starts with "TA"; therefore, the first transition is not counted for $T_{T \to A}$. By replacing the elements of the matrix with transition probability values calculated in Table 5.5, the following can be written:

$$P = \begin{pmatrix} 1/6 & 3/6 & 1/6 & 1/6 \\ 4/9 & 4/9 & 0/9 & 1/9 \\ 1/7 & 0/7 & 2/7 & 4/7 \\ 0/7 & 2/7 & 4/7 & 1/7 \end{pmatrix} = \begin{pmatrix} 0.166 & 0.5 & 0.166 & 0.166 \\ 0.444 & 0.444 & 0 & 0.111 \\ 0.142 & 0 & 0.285 & 0.571 \\ 0 & 0.285 & 0.571 & 0.142 \end{pmatrix}$$

The transition matrix (P) was built based on the sequence of observations (E). Consider that the following three nucleotides in the DNA sequence have to be predicted. Temporarily, the sequence (E) is increased with three unknowns ("???"):

$$E = \text{"TACTTCGATTTAAGCGCGGCGGCCTATATT}\underline{\text{A}}\text{???"}$$

Table 5.5 From observations to transition probabilities. The table shows the methodology of obtaining the transition probabilities between letters of sequence E. Red arrows from the diagram indicate zero transition probability in the context of sequence E. The appropriate transition matrix "*P*" for this case is shown below the diagram.

The observed sequence	Markov diagram

E = "TACTTCGATTTAAGCGCGGCG
  GCCTATATTA"

$P[A|A]$ = TACTTCGATTTAAGCGCG
  GCGGCCTATATTA = 1/6

$P[A|T]$ = TACTTCGATTTAAGCGCG
  GCGGCCTATATTA = 3/6

$P[A|G]$ = TACTTCGATTTAAGCGCG
  GCGGCCTATATTA = 1/6

$P[A|C]$ = TACTTCGATTTAAGCGCG
  GCGGCCTATATTA = 1/6

$P[T|A]$ = TACTTCGATTTAAGCGCG
  GCGGCCTATATTA = 4/9

$P[T|T]$ = TACTTCGATTTAAGCGCG
  GCGGCCTATATTA = 4/9

$P[T|G]$ = TACTTCGATTTAAGCGCG
  GCGGCCTATATTA = 0/9

$P[T|C]$ = TACTTCGATTTAAGCGCG
  GCGGCCTATATTA = 1/9

$P[G|A]$ = TACTTCGATTTAAGCGCG
  GCGGCCTATATTA = 1/7

$P[G|T]$ = TACTTCGATTTAAGCGCG
  GCGGCCTATATTA = 0/7

$P[G|G]$ = TACTTCGATTTAAGCGC
  GGCGGCCTATATTA = 2/7

$P[G|C]$ = TACTTCGATTTAAGCGCG
  GCGGCCTATATTA = 4/7

$P[C|A]$ = TACTTCGATTTAAGCGCG
  GCGGCCTATATTA = 0/7

$P[C|T]$ = TACTTCGATTTAAGCGC
  GGCGGCCTATATTA = 2/7

$P[C|G]$ = TACTTCGATTTAAGCGCG
  GCGGCCTATATTA = 4/7

$P[C|C]$ = TACTTCGATTTAAGCGCG
  GCGGCCTATATTA = 1/7

$$P = \begin{pmatrix} P[A|A] & P[A|T] & P[A|G] & P[A|C] \\ P[T|A] & P[T|T] & P[T|G] & P[T|C] \\ P[G|A] & P[G|T] & P[G|G] & P[G|C] \\ P[C|A] & P[C|T] & P[C|G] & P[C|C] \end{pmatrix}$$

Unrelated to those presented here, but interesting to note, are the recurring digital representations of some unit fractions from our transition matrix which are related to cyclic numbers. For instance:

$1/7 = 0.\underline{142857}142857142857\ldots$
$2/7 = 0.\underline{285714}285714285714\ldots$
$3/7 = 0.\underline{428571}428571428571\ldots$
$4/7 = 0.\underline{571428}571428571428\ldots$
$5/7 = 0.\underline{714285}714285714285\ldots$
$6/7 = 0.\underline{857142}857142857142\ldots$

The cyclic number phenomenon is more obvious in the case of integers. Notice that in a cyclic number, the cyclic permutations of the digits are successive multiples of the number. The following formula generates cyclic numbers using prime numbers:

$$\text{Cyclic number} = \frac{10^{p-1} - 1}{p}$$

where p is a prime number. In this case, the prime number 7 is used:

$$\frac{10^{7-1} - 1}{7} = 142857$$

However, not all prime numbers generate a cyclic number. It is estimated that only 37% of prime numbers generate a cyclic number when replaced in the formula above. As an example, the digits of 142857 shift cyclically when it is multiplied by a number n.

$142857 \times 1 = 14285\underline{7}$
$142857 \times 3 = 42857\underline{1}$
$142857 \times 2 = 28571\underline{4}$
$142857 \times 6 = 85714\underline{2}$
$142857 \times 4 = 57142\underline{8}$
$142857 \times 5 = \underline{7}14285$

where $n = 1, 3, 2, 6, 4, 5$.

The prediction rules have been previously discussed and are perhaps familiar from the examples given for two- or three-state Markov chains. Therefore, the initial state vector ($v^{(0)}$) is built according to the latest observation in the sequence (i.e., "**A**"). There are four possible states which are represented by a probability vector ($v^{(0)}$) with four components, namely:

$$v^{(0)} = [x_0 \quad y_0 \quad z_0 \quad w_0]$$

where each component in the probability vector ($v^{(0)}$) corresponds to:

$$v^{(0)} = [P[\text{Adenine}] \quad P[\text{Thymine}] \quad P[\text{Cytosine}] \quad P[\text{Guanine}]]$$

The current event is represented by the last state in the sequence (E), namely Adenine ("**A**"). Since it represents the current observation, the probability of Adenine in the initial state vector ($v^{(0)}$) is 1. The sum of the components of a probability vector makes unity ($x_0 + y_0 + z_0 + w_0 = 1$). Therefore, the other three components have zero probability:

$$v^{(0)} = [1 \quad 0 \quad 0 \quad 0]$$

Up to this point, the initial state vector and the transition matrix have been set. A new vector ($v^{(1)}$) can be obtained by using the transition matrix (P) and the probability vector ($v^{(0)}$) from above. The purpose of the new vector ($v^{(1)}$) is to indicate the state of the system when the next discrete step is made. The structure of the new vector also features four components as follows:

$$v^{(1)} = [x_1 \quad y_1 \quad z_1 \quad w_1]$$

In order to calculate this new vector ($v^{(1)}$), the initial state vector ($v^{(0)}$) is multiplied by the transition matrix (P):

$$v^{(0)} \times P = v^{(1)}$$

The coordinates of the transition matrix are used as (m_{ij}):

$$[x_0 \quad y_0 \quad z_0 \quad w_0] \begin{pmatrix} m_{11} & m_{12} & m_{13} & m_{14} \\ m_{21} & m_{22} & m_{23} & m_{24} \\ m_{31} & m_{32} & m_{33} & m_{34} \\ m_{41} & m_{42} & m_{43} & m_{44} \end{pmatrix} = [x_1 \quad y_1 \quad z_1 \quad w_1]$$

The elements of the transition matrix are replaced with the transition probability values:

$$[1 \quad 0 \quad 0 \quad 0] \begin{pmatrix} 0.166 & 0.500 & 0.166 & 0.166 \\ 0.444 & 0.444 & 0 & 0.111 \\ 0.142 & 0 & 0.285 & 0.571 \\ 0 & 0.285 & 0.571 & 0.142 \end{pmatrix} = [x_1 \quad y_1 \quad z_1 \quad w_1]$$

Similar to the example of Markov chains with three states, the colors are meant to help with the visualization of the resulting vector. Note that in the case of the initial probability vector ($v^{(0)}$), the orange color represents the first component and is associated with the "Adenine" state. The blue color represents the second component of this vector and is associated with the "Thymine" state. The gray color represents the third component of this vector and is associated with the "Cytosine" state, and the burgundy color represents the fourth element of this vector and is associated with the "Guanine" state.

The colors of each component from the resulting vector ($v^{(1)}$) have the same meaning. The transition matrix is made also of four row probability vectors representing the state transitions. In this case, the orange color is associated with the first row probability vector that represents the transitions from "Adenine" to another state (i.e., "A" to "A", "A" to "T", "A" to "C", or "A" to "G"). The blue color is associated with the second row probability vector that represents the transitions from "Thymine" to another state. The gray color is associated with the third row probability vector that represents the transitions from "Cytosine" to another state. And last, the burgundy color is associated with the fourth row probability vector that represents the transitions from "Guanine" to another state. As before, each component of the new probability vector ($v^{(1)}$) is calculated as follows:

$$x_1 = (x_0 \times m_{11}) + (y_0 \times m_{21}) + (z_0 \times m_{31}) + (w_0 \times m_{41})$$
$$y_1 = (x_0 \times m_{12}) + (y_0 \times m_{22}) + (z_0 \times m_{32}) + (w_0 \times m_{42})$$
$$z_1 = (x_0 \times m_{13}) + (y_0 \times m_{23}) + (z_0 \times m_{33}) + (w_0 \times m_{43})$$
$$w_1 = (x_0 \times m_{14}) + (y_0 \times m_{24}) + (z_0 \times m_{34}) + (w_0 \times m_{44})$$

As a general rule, in each step (k), the probability vector ($v^{(k-1)}$) that resulted from the previous step ($k-1$) is multiplied by the transition matrix (P) until the chain reaches the end. In our particular case, three steps are needed for the prediction of the next three nucleotides in the sequence (E). At the end of each step, the components of the resulting vectors are evaluated. The component with the highest probability value indicates the most likely state in which the system will be in that step. Therefore, the probability vectors are computed for each step, starting from the initial state vector ($v^{(0)}$):

Nucleotide Prediction in Step 1

$$v^{(0)} \times P = v^{(1)}$$
$$[1 \quad 0 \quad 0 \quad 0] \times P = [0.166 \quad 0.5 \quad 0.166 \quad 0.166]$$

In this case, the y_1 component is showing the highest probability value. Thus, in step 1, the system will be most likely in the "Thymine" state and the first question mark from sequence (E) is replaced with letter "T": E = "TACTTCGATTTAAGCGCGGCGGCCTATATT<u>T</u>??".

Nucleotide Prediction in Step 2

$$v^{(1)} \times P = v^{(2)}$$
$$[0.166 \quad 0.5 \quad 0.166 \quad 0.166] \times P = [0.273 \quad 0.352 \quad 0.169 \quad 0.201]$$

In this case, the y_2 component is showing the highest probability value. Thus, in step 2, the system will be most likely in the "Thymine" state and the second question mark from sequence (E) is replaced with letter "T": E = "TACTTCGATTTAAGCGCGGCGGCCTATATT<u>T</u>T?".

Nucleotide Prediction in Step 3

$$v^{(2)} \times P = v^{(3)}$$

$$[0.273 \quad 0.352 \quad 0.169 \quad 0.201] \times P = [0.225 \quad 0.350 \quad 0.208 \quad 0.209]$$

In this case, the y_3 component is showing the highest probability value. Thus, in step 3, the system will be most likely, again, in the "Thymine" state and the third question mark from sequence (E) is replaced with letter "T": E = "TACTTCGATTTAAGCGCGGCGGCCTATATT<u>A</u>TTT".

The computer code below summarizes those discussed so far on the four-state Markov chain case. First, the implementation computes the transition probabilities between letters (nucleotides) of sequence E and generates a 4×4 right stochastic matrix. In the second phase, the stochastic matrix is used for the prediction of the next three nucleotides in the sequence:

```
Dim M(1 To 4, 1 To 4) As String

Private Sub Form_Load()

Dim v(0 To 3) As Variant

Call ExtractProb("TACTTCGATTTAAGCGCGGCGGCCTATATTA")

chain = 3

v(0) = 1
v(1) = 0
v(2) = 0
v(3) = 0

For i = 1 To chain

x = (v(0)*M(1, 1)) + (v(1)*M(2, 1)) + (v(2)*M(3, 1))
    + (v(3)*M(4, 1))
y = (v(0)*M(1, 2)) + (v(1)*M(2, 2)) + (v(2)*M(3, 2))
    + (v(3)*M(4, 2))
z = (v(0)*M(1, 3)) + (v(1)*M(2, 3)) + (v(2)*M(3, 3))
    + (v(3)*M(4, 3))
w = (v(0)*M(1, 4)) + (v(1)*M(2, 4)) + (v(2)*M(3, 4))
    + (v(3)*M(4, 4))

v(0) = x
v(1) = y
```

```
v(2) = z
v(3) = w

out = Empty

For c = 0 To 3
     out = out & v(c) & "|"
Next c

MsgBox "Base(" & i & ")=[" & out & "]"

BaseBy = BaseBy & Base(v())

MsgBox BaseBy

Next i

End Sub

Function Base(ByRef v() As Variant)

For i = 0 To UBound(v)

    If v(i) > old Then
        x = v(i)
        h = i
    End If

    old = x

Next i

    If h = 0 Then n = "A"
    If h = 1 Then n = "T"
    If h = 2 Then n = "G"
    If h = 3 Then n = "C"

Base = n

End Function

Function ExtractProb(ByVal s As String)
```

```
Ea = "A"
Et = "T"
Eg = "G"
Ec = "C"

For i = 1 To 4
    For j = 1 To 4
      M(i, j) = 0
    Next j
Next i

Ta = 0
Tt = 0
Tg = 0
Tc = 0

For i = 2 To Len(s) - 1

        DI1 = Mid(s, i, 1)
        DI2 = Mid(s, i + 1, 1)

        If DI1 = Ea Then r = 1
        If DI1 = Et Then r = 2
        If DI1 = Eg Then r = 3
        If DI1 = Ec Then r = 4

        If DI2 = Ea Then c = 1
        If DI2 = Et Then c = 2
        If DI2 = Eg Then c = 3
        If DI2 = Ec Then c = 4

        M(r, c) = Val(M(r, c)) + 1

        If DI1 = Ea Then Ta = Ta + 1
        If DI1 = Et Then Tt = Tt + 1
        If DI1 = Eg Then Tg = Tg + 1
        If DI1 = Ec Then Tc = Tc + 1

Next i

For i = 1 To 4
    For j = 1 To 4
```

```
        If i = 1 Then M(i, j) = Val(M(i, j)) / Ta
        If i = 2 Then M(i, j) = Val(M(i, j)) / Tt
        If i = 3 Then M(i, j) = Val(M(i, j)) / Tg
        If i = 4 Then M(i, j) = Val(M(i, j)) / Tc
    Next j
Next i


End Function
```

```
Output:
Bases(1)=[0.166666666666667|0.500000000000|0.166666666666|
0.166666666666]
Bases(2)=[0.273809523809524|0.353174603174|0.170634920634|
0.202380952380]
Bases(3)=[0.226977828168304|0.351694381456|0.210034013605|
0.211293776769]
BasesPredicted=[TTT]
```

Supporting algorithm 9: Step-by-step prediction by using a DNA sequence. The letters that make up a DNA sequence are: "A", "T", "G", and "C". Thus, the observations present in a DNA sequence are suitable for exemplifications involving a four-state Markov machine. As before, a 4 × 4 matrix is used for counting all the combinations of pairs of letters ($D_{a \to b}$) in the DNA sequence ($D_{a \to b}$ is represented by joining two string variables, namely DI1 and DI2). In parallel, the first letter (N_a) of each pair is counted inside the DNA sequence (N_a is represented by variable DI1). Second, the transition probabilities are computed. The values from each element of the matrix are divided by their corresponding N_a. In the final phase, a probability vector is repeatedly multiplied by the new transition matrix. The vectors obtained from these repetitions show the probability of each outcome on a particular step.

5.4 Predictions by Using *n*-State Markov Chains

A transition matrix can get quite large from case to case. In the previous example, a DNA sequence has been analyzed by means of a 4 × 4 transition matrix. However, other sequences with several types of letters can produce even larger transition matrices. The analysis of proteins is a good example. Protein sequences may incorporate up to 20 types of amino acids in their structure. This indicates that a Markov chain for protein analysis may rise up to 20 states. A transition matrix for 20 states reaches 20 × 20 = 400 transitions. Moreover, in linguistic studies, a transition matrix can reach huge dimensions if the transitions between the words of a text are considered. Thus, as more states are added to a Markov chain, the number of transitions grows quadratically (one row ($i + 1$) and one column ($j + 1$) is added):

$$P = \begin{pmatrix} m_{11} & \cdots & m_{1j} \\ \vdots & \ddots & \vdots \\ m_{i1} & \cdots & m_{ij} \end{pmatrix}$$

where i represents the number of vectors (matrix rows) that make up the matrix (P) and j represents the number of components of a vector (matrix columns). When a complex analysis is needed and the state space becomes wider (as in the case of protein analysis), multiplying a probability vector by a transition matrix becomes difficult to follow in a Markov chain. So far, Markov chains with two, three, and four states have been observed. Also, for convenience, different letters have been used to represent the vector components. For instance, the four-state Markov chain has probability vectors with four components represented by different letters, namely x, y, z, and w:

$$v^{(k)} = [x_k \quad y_k \quad z_k \quad w_k]$$

where k is an index for the steps taken in the Markov chain. Thus, in order to perform the matrix multiplication with the probability vectors, the expressions below have been used:

$$x_{k+1} = (x_k \times m_{11}) + (y_k \times m_{21}) + (z_k \times m_{31}) + (w_k \times m_{41})$$
$$y_{k+1} = (x_k \times m_{12}) + (y_k \times m_{22}) + (z_k \times m_{32}) + (w_k \times m_{42})$$
$$z_{k+1} = (x_k \times m_{13}) + (y_k \times m_{23}) + (z_k \times m_{33}) + (w_k \times m_{43})$$
$$w_{k+1} = (x_k \times m_{14}) + (y_k \times m_{24}) + (z_k \times m_{34}) + (w_k \times m_{44})$$

It becomes obvious that different letters for each component cannot be used for n states. For instance, if there are more than 24 states in some special case, the letters of the alphabet are not sufficient for the representation of the vector components. Instead, a classical notation can be used:

$$v = [v_1 \quad v_2 \quad \cdots \quad v_n] \quad \sum_{i=1}^{n} v_i = 1$$

where v is a row probability vector with n components (states) and i is the index of the components. In the examples used so far, the index of a component (i) was represented by a unique letter (x, y, z, and w). The correspondence between the old notation and the new notation of the components is shown below:

$$v_1^{(k)} = x_k$$
$$v_2^{(k)} = y_k$$
$$v_3^{(k)} = z_k$$
$$v_4^{(k)} = w_k$$

The vectors of a four-state Markov chain can be written as:

$$v_i^{(k)} = \left[v_1^{(k)} \quad v_2^{(k)} \quad v_3^{(k)} \quad v_4^{(k)} \right]$$
$$\sum_{i=1}^{n=4} v_i^{(k)} = 1$$

where v is a row probability vector with n components (states), i represents the index of the components, and k represents an index of the steps (the chain level). Now the expressions can be rewritten as:

$$v_1^{(k+1)} = \left(v_1^k \times m_{11}\right) + \left(v_2^k \times m_{21}\right) + \left(v_3^k \times m_{31}\right) + \left(v_4^k \times m_{41}\right)$$
$$v_2^{(k+1)} = \left(v_1^k \times m_{12}\right) + \left(v_2^k \times m_{22}\right) + \left(v_3^k \times m_{32}\right) + \left(v_4^k \times m_{42}\right)$$
$$v_3^{(k+1)} = \left(v_1^k \times m_{13}\right) + \left(v_2^k \times m_{23}\right) + \left(v_3^k \times m_{33}\right) + \left(v_4^k \times m_{43}\right)$$
$$v_4^{(k+1)} = \left(v_1^k \times m_{14}\right) + \left(v_2^k \times m_{24}\right) + \left(v_3^k \times m_{34}\right) + \left(v_4^k \times m_{44}\right)$$

The burgundy color of the subscript indicates that iterations can be used. Since the number of rows (i) of the matrix ($m_{(row)(column)}$) is equal to the number of components (n) of the vector (v), the ith component of the new vector can be written as:

$$v_i^{(k+1)} = \sum_{i=1}^{n} v_i^{(k)} \times m_{ij}$$

The computer code below summarizes those discussed so far on the four-state Markov chain case. First, the implementation computes the transition probabilities between letters (nucleotides) of sequence E and generates a 4×4 right stochastic matrix. In the second phase, the stochastic matrix is used for the prediction of the next three nucleotides in the sequence by using the above expression:

```
Dim M(1 To 4, 1 To 4) As String

Private Sub Form_Load()

Dim v(0 To 3, 0 To 1) As Variant

Call ExtractProb("TACTTCGATTTAAGCGCGGCGGCCTATATTA")

chain = 5

v(0, 0) = 1
v(1, 0) = 0
v(2, 0) = 0
v(3, 0) = 0

v(0, 1) = 0
v(1, 1) = 0
```

```
v(2, 1) = 0
v(3, 1) = 0

For k = 1 To chain

    For i = 0 To 3
        For j = 0 To 3
            v(i, 1) = v(i, 1) + (v(j, 0) * M(j + 1, i + 1))
        Next j
    Next i

    For i = 0 To 3
        v(i, 0) = v(i, 1)
        v(i, 1) = 0
    Next i

    A = v(0, 0)
    T = v(1, 0)
    C = v(2, 0)
    G = v(3, 0)

    MsgBox "V(" & k & ")=[" & A & " | " & T & " | " & C &
" | " & G & "]"

Next k

End Sub
```

```
Output:
V(1)=[0.16666666666666|0.50000000000000|0.16666666666666|
    0.1666666666666]
V(2)=[0.27380952380952|0.35317460317460|0.17063492063492|
    0.2023809523809]
V(3)=[0.22697782816830|0.35169438145628|0.21003401360544|
    0.2112937767699]
V(4)=[0.22414311109511|0.33016717857042|0.21857865721244|
    0.2271110531220]
V(5)=[0.21532367589496|0.32370139612165|0.22958597474151|
    0.2313889532418].
```

Supporting algorithm 10: Predictions based on sequences produced by *n*-state Markov machines. This example also uses a DNA sequence as a model. However, the algorithm allows for an unlimited number of letters (observations). Previously, the vector–matrix multiplication cycle was declared manually with a range of expressions. Here, the multiplication cycle is made iteratively. For a prediction on more than four states, the matrix elements and the number of vector components can be increased to cover a new prediction requirement. Note that "ExtractProb" function is not shown. However, when the above algorithm is used, the "ExtractProb" function must be present.

5.5   Markov Chain Modeling on Measurements

In most cases, the data originate from precise measurements of different phenomena that cannot be directly interpreted with the help of a DMTC. An example in this respect may be related to the field of diabetology (medicine). Human body requires glucose for the production of energy [39]. Glucose in blood is commonly known as blood sugar. In the long run, both low glucose levels (concentrations) and elevated glucose levels can lead to serious health complications [40–43]. The body naturally regulates blood glucose levels with the help of the insulin hormone (secreted by the pancreas) [39]. If blood sugar levels are either increased or decreased compared to the observed limits seen in normal individuals, it might indicate a medical condition [40–43]. Serious fluctuations in the levels of blood sugar lead to complications such as diabetes or stroke [39, 42]. Thus, in order to help diagnose diabetes, blood tests provide precise levels of blood glucose. A single test generally provides a number between 65 and 200 which represents the level of blood glucose expressed in milligrams per deciliter (mg/dL) [44]. The results of these glycemic measurements are dependent on mealtimes. The question should be: Since mealtimes are virtually unknown events, can the patient's glycemic level be predicted for the next period? As discussed in previous chapters, Markov chains method can capture the behavior of the system. Thus, a Markov chain may indicate the patient's future blood glucose based on its previous daily behavior. Assuming that an individual does this test three times a day for a period of 2 weeks, it will obtain a series of numbers (a list of blood glucose levels). Thus, a total of 42 observations are expected (3 tests × (7 days + 7 days) = 3 × 14 = 42). Suppose that these 42 results (R) are written down on a paper in the form of continuous strings of numbers as follows:

R = 159, 82, 187, 194, 179, 115, 197, 102, 105, 104, 95, 126, 74, 143, 143, 127, 98, 70, 92, 170, 168, 182, 149, 85, 137, 100, 170, 180, 61, 177, 86, 195, 198, 182, 150, 197, 103, 103, 186, 100, 96, 196

For molding the glycemic levels in a Markovian manner, a simple Markov machine with only two states is first considered. A question that comes naturally in mind would be: *how can such strings of numbers be analyzed as a sequence of observations/events?* One approach for the use of the two-state model consists of an abstraction of the blood glucose values based on a threshold, namely: above normal or below normal (Figure 5.3a). This approach might be viewed in terms of "which is dominant". After all, a behavioral analysis takes

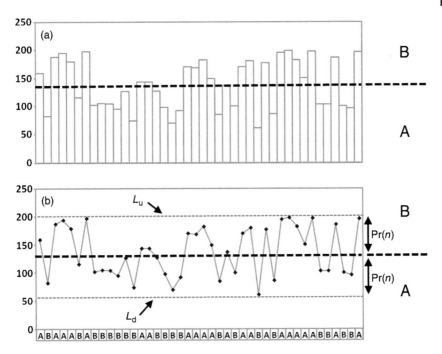

Figure 5.3 From measurements to events. (a) Glycemic values from measurements are plotted on a classic graph in the form of 42 vertical bars. The black dotted line shows the halfway between the upper limit ($L_u = 60$) and the lower limit ($L_d = 200$). (b) Shows the same glycemic values plotted as points interconnected with orange lines. Dotted lines represent the thresholds calculated for a diagram with two states ($n = 2$). The red dotted lines represent the upper limit (L_u) and the lower limit (L_d). Between these dotted lines are the state regions of size Pr(n) corresponding to each state. Each region has an associated letter. Glycemic values entering one or the other region are associated with a letter representative of a state. At the bottom of the chart, the 42 letters shown on the *x*-axis correspond to the glycemic values plotted above the letters.

into consideration the transitions between glycemic events and puts less importance on the exact blood glucose values. Thus, this threshold can reveal two types of events, namely: "A" (if glucose is below normal) and "B" (if glucose is above normal). Thus, one way to accomplish this is the division of the *y*-axis of the graph in regions of equal size (Pr). Each region will be considered a state range (Pr). However, two limits of the blood glucose have been discussed. The upper limit of glucose is represented by 200 mg/dL and the lower limit of glucose is represented by 60 mg/dL. The region of the *y*-axis that is to be divided lies between 60 and 200 mg/dL. In order to obtain the segment between 60 and 200 mg/dL, the down limit (L_d) is subtracted from the upper limit (L_u). Onward, the state range (Pr) is obtained by dividing this segment ($L_u - L_d$) by the number of states (n) used in the analysis. Thus, the state range (Pr) on the y-axis is calculated according to the following formula:

$$Pr(n) = \frac{(L_u - L_d)}{n}$$

where n represents the chosen number of states to work with, L_u represents the upper limit (the upper limit of glucose, namely 200 mg/dL), and L_d represents the down limit (the lower limit of glucose, namely 60 mg/dL). Taking the glucose example, this experiment can be analyzed with a variable number of states (n). For instance, if two states ($n = 2$) are used, the state range (Pr) is:

$$Pr(2) = \frac{(L_u - L_d)}{n} = \frac{(200 - 60)}{2} = \frac{140}{2} = 70$$

If three states ($n = 3$) are used, the state range (Pr) is:

$$Pr(3) = \frac{(L_u - L_d)}{n} = \frac{(200 - 60)}{3} = \frac{140}{3} = 46$$

If four states ($n = 4$) are used, the state range (Pr) is:

$$Pr(4) = \frac{(L_u - L_d)}{n} = \frac{(200 - 60)}{4} = \frac{140}{4} = 35$$

Up to this point, the sizes of the state regions are known. However, the borders of the state regions on the y-axis should be also determined. These boundaries act as threshold values for later use. Thus, for each state, one Pr value is added to the lower limit (L_d) value. For example, for state one, one Pr value is added to the lower limit:

$$T = L_d + Pr$$

For state two, two Pr values are added to the lower limit:

$$T = L_d + Pr + Pr$$

For the state three case, three Pr values are added to the lower limit, and so on:

$$T = L_d + Pr + Pr + Pr$$

Thus, the thresholds can be determined using the following formula:

$$T(s) = L_d + Pr \times (s - 1)$$

where s represents the state and Pr represents the state range on the chart (Figure 5.3b). These cases can be considered one by one for two states, three states, or four states:

(a) If two states are used

$$T(1) = L_d + \text{Pr} \times (s - 1) = 60 + 35 \times (1 - 1) = 60$$

$$T(2) = L_d + \text{Pr} \times (s - 1) = 60 + 35 \times (2 - 1) = 130$$

(b) If three states are used

$$T(1) = L_d + \text{Pr} \times (s - 1) = 60 + 35 \times (1 - 1) = 60$$

$$T(2) = L_d + \text{Pr} \times (s - 1) = 60 + 35 \times (2 - 1) = 107$$

$$T(3) = L_d + \text{Pr} \times (s - 1) = 60 + 35 \times (3 - 1) = 153$$

(c) If four states are used

$$T(1) = L_d + \text{Pr} \times (s - 1) = 60 + 35 \times (1 - 1) = 60$$

$$T(2) = L_d + \text{Pr} \times (s - 1) = 60 + 35 \times (2 - 1) = 95$$

$$T(3) = L_d + \text{Pr} \times (s - 1) = 60 + 35 \times (3 - 1) = 130$$

$$T(4) = L_d + \text{Pr} \times (s - 1) = 60 + 35 \times (4 - 1) = 165$$

The next step consists in converting the glucose values to observations. Visually, the type of event can be determined through a simple inspection made on the charts (Figure 5.4). However, when a glycemic value is provided, it must be associated automatically with one of the states. For instance, if a total of four states ($n = 4$) are used, this numerical association may be: 0 corresponds to letter "A", 1 corresponds to letter "B", 2 corresponds to letter "C", and 3 corresponds to letter "D". In order to reach this association, a formula is needed. To construct such a formula, two steps are required. First, if the lower limit (L_d) is subtracted from the glycemic value (Input), a range is obtained. Second, this range (Input $- L_d$) is then divided by the state range (Pr). The number obtained from this division ($S(\text{Input})$) indicates the state associated with the glycemic value. The below formula determines the state depending on the input value:

$$S(\text{Input}) = \frac{(\text{Input} - L_d)}{\text{Pr}(n)}$$

The input can be verified by:

$$\text{Input} = S(\text{Input}) \times \text{Pr}(n) + L_d$$

where Input represents the glycemic value, L_d represents the lower limit, $\text{Pr}(n)$ represents the state range, and n represents the number of states used. Thus, the $S(\text{Input})$ values that enter these regions are associated with that particular

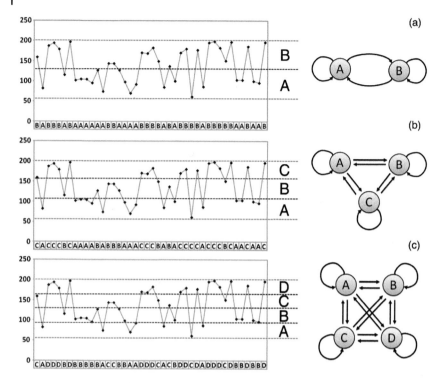

Figure 5.4 From measurements to events by using a variable number of states. (a) Shows the glycemic values divided into two state regions of size Pr(n), where $n = 2$ (uses a two-state diagram). (b) Shows the glycemic values divided into three state regions of size Pr(n), where $n = 3$ (uses a three-state diagram). (c) Shows the glycemic values divided into four state regions of size Pr(n), where $n = 4$ (uses a four-state diagram). Dotted lines represent the thresholds calculated for a state diagram. In each case, the red dotted lines represent the upper limit (L_u) and the lower limit (L_d). Between these dotted lines are the state regions of size Pr(n). Depending on the number of states, each region has an associated letter. Glycemic values entering these regions are associated with a letter. Therefore, the letter represents the final observation for a glycemic value. At the bottom of each chart, the 42 letters (x-axis) correspond to the glycemic values plotted above.

state. For example, the first glycemic value can be considered from the R string provided above, namely 159. Thus, this glycemic value can be associated with one of the states depending on the total number of states used (n):

(a) For two states

$$S(159) = \frac{(\text{Input} - L_d)}{\text{Pr}(2)} = \frac{(159 - 60)}{70} = 1.41 = 1$$

The interval between 0 and 0.999 belongs to state "A" and the interval between 1 and 1.999 belongs to state "B". By *rounding down* any value provided by the above formula, the number of the states can be obtained as an integer value (i.e., 0.999 is 0 or 1.999 is 1). Thus, each integer corresponds to a letter. In this particular case: 0 corresponds to letter "A" and 1 corresponds to letter "B". For instance, by applying the above formula for each value in R, it can be seen that 159 falls in the region of state "B", 82 falls in the region of "A", 187 falls in the region of "B", and so on until the last value in R (Figure 5.3b). If this reasoning is applied to all the values in R, then the string of observations will translate into:

Obs = "BABBBABAAAAAABBAAAABBBBABABB
BBABBBBBAABAAB"

Thus, after getting the entire series of observations as shown above, the glycemic events can be studied in a Markovian manner as described in Supporting algorithm 7.

(b) For three states
The only change that occurs compared to the previous case is represented by $\Pr(n)$, because the total number of states (n) grows from two states ($n = 2$) to three states ($n = 3$):

$$S\,(159) = \frac{\left(\text{Input} - L_{\text{d}}\right)}{\Pr(3)} = \frac{(159 - 60)}{46} = 2.12 = 2$$

As before, the interval between 0 and 0.999 belongs to state "A", the interval between 1 and 1.999 belongs to state "B", and the interval between 2 and 2.999 belongs to state "C". The value (2.12) provided by function S falls in this case in the region of another state, namely state "C" (0 corresponds to letter "A", 1 corresponds to letter "B", and 2 corresponds to letter "C"). If this reasoning is applied to all the values in R, then the string of observations translates into:

Obs = "CACCCBCAAAABABBBAAACCCBABACCCCACC
CBCAACAAC"

Since the series of observations is known, the glycemic events can be studied based on a diagram with three states (Supporting algorithm 8).

(c) For four states
If four states are used, the change that occurs is represented again only by $\Pr(n)$:

$$S\,(159) = \frac{\left(\text{Input} - L_{\text{d}}\right)}{\Pr(4)} = \frac{(159 - 60)}{35} = 2.82 = 2$$

This time there are four intervals: the interval between 0 and 0.999 is associated with state "A", the interval between 1 and 1.999 is associated with state "B", the interval between 2 and 2.999 is associated with state "C", and the interval between 3 and 3.999 is associated with state "D". Following each value in *R*, the string of observations translates into:

Obs = "CADDDBDBBBBBACCBBAADDDCACBDDCDA
        DDDCDBBDBBD"

Thus, the glycemic events from Obs can be studied by using Supporting algorithm 9. The same example, based on blood sugar values, can be analyzed using various states. Here, diagrams of two, three, and four states have been used for illustration. The implementation below uses the above formulas for Pr and *S* in a four states (*n* = 4) context to produce the Obs sequence of observations:

```
Private Sub Form_Load()
Dim Inp() As String
Dim R As String

R = "159,82,187,194,179,115,197,102,105,104,95,126,74,143,143,
127,98," & _ "70,92,170,168,182,149,85,137,100,170,180,61,177,
86,195,198,182,150," & _ "197,103,103,186,100,96,196"

Inp = Split(R, ",")
Lu = 200
Ld = 60
n = 4
Pr = (Lu - Ld) / n

For i = 0 To UBound(Inp)
    s = (Inp(i) - Ld) / Pr
    s = Split(s, ".")(0)

    If s = 0 Then l = "A"
    If s = 1 Then l = "B"
    If s = 2 Then l = "C"
    If s = 3 Then l = "D"

    Obs = Obs & l
    Reg = Reg & s & ","
Next i

MsgBox "Reg=" & Reg & vbCrLf & "Obs=" & Obs & vbCrLf
End Sub
```

```
Output:
Reg=2,0,3,3,3,1,3,1,1,1,1,1,0,2,2,1,1,0,0,3,3,3,2,0,2,1,3,3,2,3,
0,3,3,3,2,3,1,1,3,1,1,3,
Obs=CADDDBDBBBBBACCBBAADDDCACBDDCDADDDCDBBDBBD
```

Supporting algorithm 11: The conversion of measurements to states. A range of values is divided into four equal regions. Each region corresponds to a state: "A", "B", "C", and "D". The numeric values are associated with a representative letter based on their position over the regions. Thus, the initial values are listed as letters (observations).

In the first instance, the constant parameters are declared, such as *Lu*, *Ld*, and *n*. Also, Pr is calculated according to the total number of states *n*. In the next step, *S* is calculated for each value in *R*. At each iteration, the value provided by *S* is rounded down and added to the Reg variable. In the same step, the letter corresponding to the value provided by *S* is added to the Obs variable. After the iterative process, both Reg and Obs variables are displayed in the main output.

6

Absorbing Markov Chains

6.1 Introduction

A classic illustration for understanding the absorbing state of a Markov chain consists in a simplistic timeline simulation of a living creature [45, 46]. Undoubtedly, a creature can be described by using many states. However, in this example, just three states are chosen for convenience: "sick" (state "A"), "healthy" (state "B"), and "dead" (state "C"). Note that these states represent events which exclude each other. To oversimplify matters, an additional assumption in this model is that there are no predators or accidental deaths (in this way a direct route from "healthy" (state "B") to "dead" (state "C") it is avoided). The creature can have a transition from "sick" (state "A") to "dead" (state "C"), but never vice versa. Thus, the "dead" (state "C") state is the absorbing state for this chain. Depending on the organism and the environmental conditions, the likelihood of moving from "healthy" to "sick" or from "sick" to "healthy" or from "sick" to "dead" may carry different transition probabilities. In this case, the main assumption is that the transition probabilities between states are unknown. Thus, in this chapter, by considering this example further, the timeline simulation of a living creature is modeled as in previous cases by analyzing a special configuration of jars and balls.

6.2 The Absorbing State

A Markov chain is irreducible if all states communicate with each other. So far the main theme was about irreducible Markov chains. However, consider three states: state "A", state "B", and state "C" (Figure 6.1). Three arrows leave state "A", two arrows leave state "B", and only one arrow leaves state "C" (Figure 6.1a). Note that these arrows do not have any probabilities attached. What can be said about this system? Consider each state as a unity, namely 1. The number of arrows that depart from a state divides this unity. For instance, if three arrows depart from state "A", then the probability of leaving state "A" through

Markov Chains: From Theory to Implementation and Experimentation, First Edition. Paul A. Gagniuc.
© 2017 John Wiley & Sons, Inc. Published 2017 by John Wiley & Sons, Inc.
Companion website: www.wiley.com/go/gagniuc/markovchains

(a)

(b)

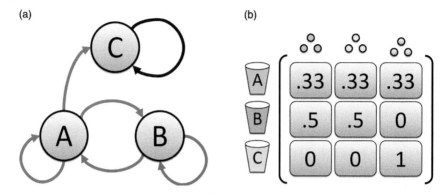

Figure 6.1 Absorbing state example. (a) The state diagram of a chain with three states. (b) The matrix lines represent the inside of the orange, blue and gray jars. The matrix columns represent the proportion of balls of different colors, namely orange balls, blue balls, and gray balls.

one of the arrows is the number of arrows that divide the unity, namely 1/3 ($P[A|$anywhere$] = 1/3 = 0.33$). Two arrows depart from state "B", then the probability of leaving state "B" through one of the arrows is 1/2 ($P[B|$anywhere$] = 1/2 = 0.5$). However, only one arrow departs from state "C", thus the probability of leaving state "C" through that arrow is 1/1 ($P[C|$anywhere$] = 1/1 = 1$). Now, the diagram shows that each state receives two arrows (Figure 6.1a). State "A" is sending arrows to state "B" and "C", and another arrow to itself. State "B" is sending an arrow to state "A" and another arrow to itself. However, note that the only arrow that goes from state "C" also returns to state "C", making a closed loop (Figure 6.1a). Consider a concrete example. Suppose that states "A", "B", and "C" are jars of different colors filled with balls of different colors. First, a transition matrix can be constructed according to the diagram configuration (Figures 6.1a and 6.1b). Assume that each row of the transition matrix represents one jar of a certain color: jar "A" (orange), jar "B" (blue), and jar "C" (gray). Each column represents the proportions of balls present in each jar. The first column represents the proportion of orange balls in each jar, the second column represents the proportion of blue balls in each jar, and the third column represents the proportion of gray balls in each jar. The first row in the matrix shows that the orange jar "A" contains an equal proportion of balls of each color, namely 33% orange balls, 33% blue balls, and 33% gray balls. The blue jar "B" contains 50% orange balls and 50% blue balls, and the gray jar "C" contains only gray balls. Assume that a series of draws are made respecting the chain rule. In the chain rule, the color of the ball indicates the color of the jar from which the next draw is made. As it was mentioned above, the gray jar "C" contains only gray balls. If draws begin from the gray jar "C", then all future draws are made from the gray jar "C". Eventually draws will reach jar "C" even if the first ball is drawn from the orange jar "A" or the blue jar "B". In this case, the gray jar "C" is considered an absorbing state. Note that the values on the main diagonal (m_{11} to m_{ij}) of the matrix represent transitions from a state to the same state. If any of the values on the main diagonal of the matrix is equal to 1, then the system contains an absorbing state. Nevertheless, what could be the behavior of such a system?! Whether draws are started from state "A" or state "B", the system will eventually reach state "C". But what is the natural path of the machine to equilibrium (to state "C")? The computer code implementation from below predicts the behavior of the three states diagram found in Figure 6.1a. The first difference from the previous implementation (Supporting algorithm 10) example consists of a different number of states, namely three states instead of four states. The second difference compared to the previous implementation (Supporting algorithm 10) is that the transition probabilities are filled in manually inside the matrix and not automatically (as it was the case with the "Extract-Prob" function). The new format of the transition matrix (Jar) is: "Jar(line, column) = probability value", where the lines represent jars (jar "A" = 1, jar "B" = 2, and jar "C" = 3) and the columns represent types of balls (orange balls "A" = 1, blue balls "B" = 2, gray balls "C" = 3). For instance, in order to indicate the proportion of orange balls "A" inside jar "B", the statement can be written as: Jar(2, 1) = 0.5.

```
Dim Jar(1 To 3, 1 To 3) As String

Private Sub Form_Load()
Dim v(0 To 2, 0 To 1) As Variant

Jar(1, 1) = 0.33
Jar(1, 2) = 0.33
Jar(1, 3) = 0.33

Jar(2, 1) = 0.5
Jar(2, 2) = 0.5
Jar(2, 3) = 0

Jar(3, 1) = 0
Jar(3, 2) = 0
Jar(3, 3) = 1

chain = 5

v(0, 0) = 1
v(1, 0) = 0
```

```
v(2, 0) = 0

v(0, 1) = 0
v(1, 1) = 0
v(2, 1) = 0

For k = 1 To chain

    For i = 0 To 2
        For j = 0 To 2
            v(i, 1) = v(i, 1) + (v(j, 0) * Jar(j + 1, i + 1))
        Next j
    Next i

    For i = 0 To 2
        v(i, 0) = v(i, 1)
        v(i, 1) = 0
    Next i

    A = v(0, 0)
    B = v(1, 0)
    C = v(2, 0)

    MsgBox "Step(" & k & ")=[" & A & " | " & B & " | " & C & "]"
Next k
End Sub
```

```
Output:
Step(1)=[0.333 | 0.333 | 0.333]
Step(2)=[0.278 | 0.278 | 0.444]
Step(3)=[0.231 | 0.231 | 0.537]
Step(4)=[0.193 | 0.193 | 0.614]
Step(5)=[0.161 | 0.161 | 0.678]
```

Supporting algorithm 12. Prediction based on a 3 × 3 transition matrix. Known transition probability values are directly used from a transition matrix for highlighting the behavior of an absorbing Markov chain.

The convergence toward the steady-state vector represents the natural path of the machine to equilibrium. In our example from above, the prediction has been made for five steps. Even in this short interval of five steps, a trend begins to take shape (Supporting algorithm 12). The probability that the system will be in state "C" grows after every step, whereas the probability that the system

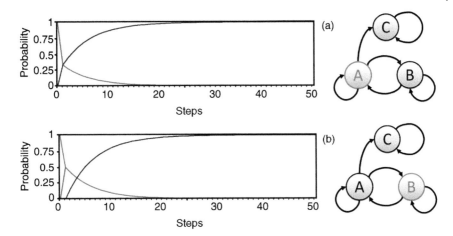

Figure 6.2 The convergence path of an absorbing Markov chain. (a) Convergence path if the system starts in state "A" (b) Convergence path if the system starts in state "B".

will be in state "A" or "B" decreases synchronously after every step. Since the system contains an absorbing state, one of the components of the stationary vector will have a value of 1 and the rest of the components a value of 0, such as: $v = [0, 0, 1]$. However, an interesting strategy for understanding this three-state system is to observe how it will behave until it reaches the stationary vector. If the vector components are plotted on a graph for 50 steps, an overview for the long-run path (convergence toward the steady-state vector) of the system can be observed (Figure 6.2). Here, the values of the vector components enter a smooth path toward the steady-state vector starting after the first step (Figure 6.2). If the system is started from jar "A", then the probability of reaching jar "B", jar "C", or returning to jar "A" in the next step is 0.33. However, if draws begin from jar "B", in the next step, there are equal chances to move to jar "A" or back into jar "B". Following the first step, it can be seen that there is an equal probability of being in state (jar) "A" or "B", regardless of the starting state. Also, both charts show that in the range of just 50 steps, the system reaches state "C" regardless of the starting state (Figures 6.2a and 6.2b).

7

The Average Time Spent in Each State

7.1 Introduction

In recent decades, Markov chains provided great solutions for many technological applications [47]. The main focus so far has been linked to simulation-based
experiments related to jars and balls. As shown in previous chapters, many situations related to natural processes can mold on the examples associated to jars
and balls. Once the transition probabilities are known for a series of observations, two fundamental questions related to a Markov chain can arise: (1) What
is the percentage of ball types in each jar? (2) On average, how often is a jar
visited if the number of draws tends to infinity (if the machine works continu-
ously)? Therefore, this chapter is dedicated to these two questions.

7.2 The Proportion of Balls in the System

Probability matrices are made of probability vectors. Each state is represented
by a probability vector. The components of a probability vector sum up to 1.
Thus, if all the values from the elements of the matrix are summed, the result is
also an integer. This integer represents the number of states in the system (q).
The columns of a transition matrix (P) indicate the proportion of a particular
type of ball that is present throughout the system. First, the proportion of a
particular type of ball from each state can be expressed as the sum over i:

$$h(s) = \sum_{i=1}^{n} P_{i,s}$$

where h is the sum of the elements from the column (s) of the matrix (P) corresponding to a state, and i represents an index for the elements present on the
column (s). The total number of elements from the column is represented by n.

Markov Chains: From Theory to Implementation and Experimentation, First Edition. Paul A. Gagniuc.
© 2017 John Wiley & Sons, Inc. Published 2017 by John Wiley & Sons, Inc.
Companion website: www.wiley.com/go/gagniuc/markovchains

Thus, the proportion of a particular type of ball that is present throughout the system can be expressed in percentages as:

$$b(s) = h(s) \times \frac{100}{q}$$

where $b(s)$ represents the proportion of a particular type of ball (s) that is present throughout the system, and q represents the total number of states. This method works for all state diagrams, including those with absorbing states. Furthermore, note that n represents the number of vectors in the right stochastic matrix (or the number of elements from the column) and q is the number of states in the system. Since a state has an associated vector, $n = q$. Thus, a shorter version can be written for $b(s)$, where n is replaced by q:

$$b(s) = \left(\sum_{i=1}^{q} m_{is} \right) \times \frac{100}{q}$$

In other words, the average of the values from the columns of the transition matrix (P) indicates the proportion of balls in the system. One may be inclined to believe that the proportion of balls throughout the system partially indicates the time spent in each state, but such an assumption does not provide an accurate prediction for the average time spent in a state and in some cases, it can provide erroneous values. For an accurate prediction of the average time spent in a state, a minimum "vision" of the future events is needed (please see the next chapter).

7.3 The Average Time Spent in A Particular State

When diagrams become more complex, our intuition over the stochastic processes is prone to erroneous beliefs. Thus, for finding the average time spent in a particular state, a more convenient approach is needed. In a first step, the transition matrix (P) is multiplied by itself ($P \times P = P^2$). Note that square matrices can be multiplied by themselves repeatedly because they have the same number of rows and columns. This repeated multiplication can be described as the power of the matrix. For instance, an $n \times n$ matrix P raised to a positive integer k is defined as:

$$P^k = \underbrace{PPP \dots P}_{k \text{ steps}}$$

For convenience, the resulting matrix (P^2) will be noted as m. Note that the resulting matrix P^2 (noted as m) represents two steps in the future of this system. Thus, the following formula is used in order to obtain each element of row i and column j in the resulting matrix m, where $m = P \times P$:

$$m_{ij} = \sum_{k=1}^{q} P_{i,k} \times P_{k,j} = (P_{1,k} \times P_{k,1}) + (P_{2,k} \times P_{k,2}) + \cdots + (P_{i,q} \times P_{q,j})$$

Thus, each m_{ij} entry is given by multiplying the entries $P_{i,k}$ (across row i of P) by the entries of the same square matrix $P_{k,j}$ (across column j of P), for $k = 1, 2, \ldots,$ q, and summing the results over k. The formula used previously for finding the proportion of balls throughout the system is now adapted for calculating the average time spent in a particular state (s). Transition matrix m is the result of multiplying matrix P with itself. Therefore, this adaptation consists in using transition matrix m instead of transition matrix (P), as before. Thus, the average time spent in a state (s) can be expressed in percentages as:

$$t(s) = \left(\sum_{i=1}^{q} m_{is} \right) \times \frac{100}{q}$$

where $t(s)$ is the average time spent in a state (s), and q represents the total number of states. As a last remark, the average of the values from the columns of the matrix (m) indicates the average time spent in a state (s) (values expressed in decimal). Note that this method can be valid for diagrams with absorbing states *only if* matrix (m) is the result of repeated multiplications of matrix (P) with itself until a stationary position is reached. For a greater accuracy on estimating the average time spent in a state (s), the matrix (P) can be multiplied by itself several times $(P \times P \times \ldots \times P = m)$. However, the minimum number of multiplications which are needed to provide an acceptable result for the majority of diagram configurations is $P \times P = m$.

7.4 Exemplification of the Average Time and Proportions

The above formulas can be used for calculating the proportion of ball types that are present throughout the system and the average time spent in each state/jar. At every discrete step, the system is in one of the states. The average time indicates how often the system will be caught in a certain state from a total number of discrete steps. A concrete example can be given (Figure 7.1a). The previous state diagram (Figure 6.1) can be modified and reused. Thus, the arrow that returns to state "C" can be redirected toward state "B", thereby creating a system without absorbing states (Figure 7.1a). Note that the probability of each arrow remained the same as in Figure 6.1. Note also that compared to the previous example (Figure 6.1), the new transition matrix (P) shows a single difference, namely a switch of the values between $P_{i,j}$ and $P_{i,j-1}$ (Figure 6.1b and Figure 7.1b).

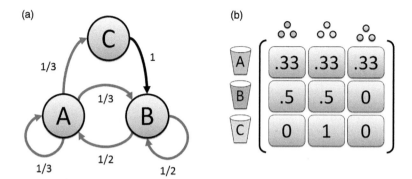

Figure 7.1 Example of state diagram. (a) The state diagram with associated probabilities. (b) The matrix lines represent the inside of the orange, blue, and gray jars. The matrix columns represent the proportion of balls of different colors over the system.

7.4.1 The Proportion of Ball Types

For this case, the transition matrix (P) derived from the state diagram is used to find the proportion of ball types ($b(s)$) that are present throughout the system (in all the jars):

$$P = \begin{pmatrix} 0.33 & 0.33 & 0.33 \\ 0.50 & 0.50 & 0.00 \\ 0.00 & 1.00 & 0.00 \end{pmatrix}$$

Furthermore, since the present case consists of three states, $q = 3$. Thus, for the proportion of orange balls, type "A", the first column is used ($s = 1$):

$$b(1) = \left(\sum_{i=1}^{3} P_{i,1} \right) \times \frac{100}{3} = (0.33 + 0.5 + 0) \times \frac{100}{3} = {\sim}28\%$$

For the proportion of blue balls type "B", the second column is used ($s = 2$):

$$b(2) = \left(\sum_{i=1}^{3} P_{i,2} \right) \times \frac{100}{3} = (0.33 + 0.5 + 1) \times \frac{100}{3} = {\sim}61\%$$

For the proportion of gray balls type "C", the third column is used ($s = 3$):

$$b(3) = \left(\sum_{i=1}^{3} P_{i,3} \right) \times \frac{100}{3} = (0.33 + 0 + 0) \times \frac{100}{3} = {\sim}11\%$$

Therefore, the system contains around 28% balls type "A", 61% balls type "B", and 11% balls type "C".

7.4.2 The Average Time Spent in a State

The average time spent in each state indicates how a system will behave in the long term. This is crucial for any type of prediction. Two steps are followed in order to determine the average time spent in a state. First, the transition matrix (P) is multiplied by itself. Thus, the following formula is used in order to obtain each element of row i and column j in the resulting matrix m:

$$m_{ij} = \sum_{k=1}^{q} P_{i,k} \times P_{k,j} = (P_{1,k} \times P_{k,1}) + (P_{2,k} \times P_{k,2}) + \cdots + (P_{i,q} \times P_{q,j})$$

where each m_{ij} entry is given by multiplying the entries $P_{i,k}$ (across row i of P) by the entries of the same square matrix $P_{k,j}$ (across column j of P). Note that the resulting matrix m represents two steps into the future of this three-state system (Figure 7.1a).

Note: Here the multiplication of two square matrices is used ($P \times P = m$):

$$m_{ij} = \begin{pmatrix} 0.33 & 0.33 & 0.33 \\ 0.50 & 0.50 & 0.00 \\ 0.00 & 1.00 & 0.00 \end{pmatrix} \times \begin{pmatrix} 0.33 & 0.33 & 0.33 \\ 0.50 & 0.50 & 0.00 \\ 0.00 & 1.00 & 0.00 \end{pmatrix} = \begin{pmatrix} m_{11} & m_{12} & m_{13} \\ m_{21} & m_{22} & m_{23} \\ m_{31} & m_{32} & m_{33} \end{pmatrix}$$

The purpose of the above expression is to calculate the values for all the elements of matrix m. For instance, in order to calculate the element in the first row ($i = 1$) and first column ($j = 1$) of m, the following applies:

$$m_{11} = \sum_{k=1}^{3} P_{1,k} \times P_{k,1} = (P_{1,1} \times P_{1,1}) + (P_{1,2} \times P_{2,1}) + (P_{1,3} \times P_{3,1})$$

$$m_{11} = (0.33 \times 0.33) + (0.5 \times 0.33) + (0 \times 0.33) = 0.3$$

The first element of the matrix m is found:

$$m_{ij} = \begin{pmatrix} 0.3 & m_{12} & m_{13} \\ m_{21} & m_{22} & m_{23} \\ m_{31} & m_{32} & m_{33} \end{pmatrix}$$

Next, another element may be calculated, for instance, the element in the second row ($i = 2$) and first column ($j = 1$) of m:

$$m_{21} = \sum_{k=1}^{3} P_{2,k} \times P_{k,1} = (P_{2,1} \times P_{1,1}) + (P_{2,2} \times P_{2,1}) + (P_{2,3} \times P_{3,1})$$

$$m_{21} = (0.33 \times 0.5) + (0.5 \times 0.5) + (0 \times 0) = 0.4$$

Thus, a second element of the matrix m is found:

$$m_{ij} = \begin{pmatrix} 0.3 & m_{12} & m_{13} \\ 0.4 & m_{22} & m_{23} \\ m_{31} & m_{32} & m_{33} \end{pmatrix}$$

This rule continues in this manner until all the elements of matrix m are filled with values.

By continuing the above example, all the elements of matrix m are filled with values:

$$m_{ij} = \begin{pmatrix} 0.33 & 0.33 & 0.33 \\ 0.50 & 0.50 & 0.00 \\ 0.00 & 1.00 & 0.00 \end{pmatrix} \times \begin{pmatrix} 0.33 & 0.33 & 0.33 \\ 0.50 & 0.50 & 0.00 \\ 0.00 & 1.00 & 0.00 \end{pmatrix} = \begin{pmatrix} 0.3 & 0.6 & 0.1 \\ 0.4 & 0.4 & 0.2 \\ 0.5 & 0.5 & 0.0 \end{pmatrix}$$

In the second stage, matrix m can be used to find the average time ($t(s)$) spent in each state (or jar). Thus, for the average time spent in state "A" the first column is used ($s = 1$):

$$t(1) = \left(\sum_{i=1}^{3} m_{i1} \right) \times \frac{100}{3} = (0.3 + 0.4 + 0.5) \times \frac{100}{3} = {\sim}40\%$$

For the average time spent in state "B", the second column is used ($s = 2$):

$$t(2) = \left(\sum_{i=1}^{3} m_{i2} \right) \times \frac{100}{3} = (0.6 + 0.4 + 5) \times \frac{100}{3} = {\sim}51\%$$

For the average time spent in state "C", the third column is used ($s = 3$):

$$t(3) = \left(\sum_{i=1}^{3} m_{i3} \right) \times \frac{100}{3} = (0.1 + 0.2 + 0) \times \frac{100}{3} = {\sim}9\%$$

Therefore, about 40% of the time draws will be made from jar "A", 51% of the time draws will be made from jar "B", and 9% of the time draws will be made from jar "C". By comparing the data obtained from the two methods, it can be noted that the proportion of balls in the system is not representative for the average time spent in a state (Table 7.1). The prediction method for the average time spent in a particular state of the system provides a very close estimate of the data derived from the experiments. The experimental results are discussed in detail in the simulation of n-state diagrams. The average time spent in each state does not necessarily relate to the stationary vector. However, this particular system has a stationary vector which relates to the average time. The same as before, the strategy for understanding this three-state system is to observe how it will behave until it reaches the stationary vector. If the vector components

Table 7.1 Comparison between global values. The proportion of balls ($b(s)$) throughout the system and the average time spent in each state ($t(s)$).

Method	State A	State B	State C
$t(s)$	40%	51%	9%
$b(s)$	28%	61%	11%

are plotted on a graph for 50 steps, an overview of the convergence toward the steady-state vector of this particular system can be observed (Figure 7.2). In the three panels of Figure 7.2, a series of lines indicate the probability of outcomes for each future discrete step made by the machine. Thus, the orange lines show the probability that the system will be in state "A". Also, the blue lines show the probability that the system will be in state "B" and the gray lines show the probability that the system will be in state "C". Notice that the values of the

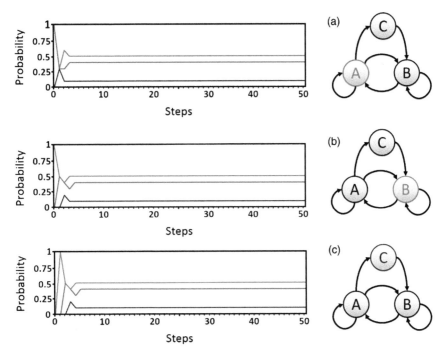

Figure 7.2 The convergence path of a three-state Markov chain. (a) Convergence path if the system starts in state "A"; (b) the convergence path if the system starts in state "B"; and (c) the convergence path if the system starts in state "C".

vector components reach the steady-state vector after just four or five steps (Figure 7.2). If the system is started from jar "A", then the probability of reaching jar "B", jar "C", or returning to jar "A" in the next step is 0.33 (Figure 7.2a). However, if draws begin from jar "B", in the next step, there are equal chances of moving to jar "A" or back into jar "B" (Figure 7.2b). Of course, if draws begin from jar "C", then there is certainty that the next draw is made from jar "B". After five steps, there is a stable probability for each outcome regardless of the starting state, namely: the probability to capture the system at any moment in state (jar) "A" ($P[A] = 0.4$), in state "B" ($P[B] = 0.51$), or in state "C" ($P[C] = 0.09$).

8

Discussions on Different Configurations of Chains

8.1 Introduction

By modifying the previous examples from Chapter 4 and Chapter 5, some inter-esting configurations of state diagrams may be further considered (Table 8.1). Most of the time, the majority of problems can be solved using Markov chains with two states. However, different configurations of state diagrams provide solutions for different problems in various fields. If this association with jars and balls is continued, the interpretation of the state diagrams from Table 8.1 can be made in an intuitive manner as the examples shown in Figure 6.1 and Figure 7.1. Table 8.1 shows 13 state diagram models. Cases A and B represent an introduction and each uses a system with two states. Cases C and D illustrate state diagrams of systems with three states. Next, state diagrams of systems with four states are shown from case E to case M. Cases H and K explain the notion of classes. Examples from cases L and M bring into focus the absorbing states con-figuration. For each case, the average time spent in each state is predicted. Also, the long-term behavior of each configuration is observed and discussed. Up to this point, the implementation for a behavior analysis has been made based on a 3×3 transition matrix (Supporting algorithm 12). However, since Table 8.1 contains various diagram configurations with a different number of states, an extension of the method is presented (Supporting algorithm 13). This exten-sion of the method allows the analysis of systems with two states, three states, or four states. Thus, the computer code implementation below may predict the behavior of any state diagram model comprising two to four states.

Markov Chains: From Theory to Implementation and Experimentation, First Edition. Paul A. Gagniuc.
© 2017 John Wiley & Sons, Inc. Published 2017 by John Wiley & Sons, Inc.
Companion website: www.wiley.com/go/gagniuc/markovchains

Table 8.1 Examples of state diagrams with different configurations. The first column in the table represents the case label, the second column represents the state diagram, and the third column shows the transition matrix associated with the state diagram. Dark gray cells from each transition matrix represent cells that contain values greater than zero. Notice the pattern of values on each transition matrix.

Case	The state diagram	Transition matrix
A		$\begin{array}{c} \quad A \quad B \\ \begin{array}{c} A \\ B \end{array} \begin{pmatrix} 0 & 1 \\ 1 & 0 \end{pmatrix} \end{array}$
B		$\begin{array}{c} \quad A \quad B \\ \begin{array}{c} A \\ B \end{array} \begin{pmatrix} 0 & 1 \\ 0.5 & 0.5 \end{pmatrix} \end{array}$
C		$\begin{array}{c} \quad A \quad B \quad C \\ \begin{array}{c} A \\ B \\ C \end{array} \begin{pmatrix} 0 & 1 & 0 \\ 0 & 0 & 1 \\ 1 & 0 & 0 \end{pmatrix} \end{array}$
D		$\begin{array}{c} \quad A \quad B \quad C \\ \begin{array}{c} A \\ B \\ C \end{array} \begin{pmatrix} 0 & 0.5 & 0.5 \\ 0 & 0.5 & 0.5 \\ 1 & 0 & 0 \end{pmatrix} \end{array}$
E		$\begin{array}{c} \quad A \quad B \quad C \quad D \\ \begin{array}{c} A \\ B \\ C \\ D \end{array} \begin{pmatrix} 0 & 1 & 0 & 0 \\ 0.5 & 0 & 0.5 & 0 \\ 0 & 0.5 & 0 & 0.5 \\ 0 & 0 & 1 & 0 \end{pmatrix} \end{array}$

Table 8.1 (*Continued*)

Case	The state diagram	Transition matrix

F

$$\begin{array}{c c c c c} & A & B & C & D \\ A & 0 & 0.5 & 0.5 & 0 \\ B & 1 & 0 & 0 & 0 \\ C & 0 & 0 & 0 & 1 \\ D & 0 & 0.5 & 0.5 & 0 \end{array}$$

G

$$\begin{array}{c c c c c} & A & B & C & D \\ A & 0 & 1 & 0 & 0 \\ B & .33 & 0 & .33 & .33 \\ C & 0 & 1 & 0 & 0 \\ D & 0 & 0 & 1 & 0 \end{array}$$

H

$$\begin{array}{c c c c c} & A & B & C & D \\ A & 0 & 1 & 0 & 0 \\ B & 0 & 0 & 1 & 0 \\ C & 0 & 0.5 & 0 & 0.5 \\ D & 1 & 0 & 0 & 0 \end{array}$$

I

$$\begin{array}{c c c c c} & A & B & C & D \\ A & 0 & .33 & .33 & .33 \\ B & 0 & 0 & 0 & 1 \\ C & 0 & 0 & 0 & 1 \\ D & 0 & 0 & 1 & 0 \end{array}$$

(*continued*)

Table 8.1 (*Continued*)

Case	The state diagram	Transition matrix
J		
K		
L		
M		

```
Dim Jar(1 To 4, 1 To 4) As String
Private Sub Form_Load()

Dim v(0 To 3, 0 To 1) As Variant

Jar(1, 1) = 1
Jar(1, 2) = 0
Jar(1, 3) = 0
Jar(1, 4) = 0

Jar(2, 1) = 0.5
Jar(2, 2) = 0
Jar(2, 3) = 0.5
Jar(2, 4) = 0

Jar(3, 1) = 0
Jar(3, 2) = 0.5
Jar(3, 3) = 0
Jar(3, 4) = 0.5

Jar(4, 1) = 0
Jar(4, 2) = 0
Jar(4, 3) = 1
Jar(4, 4) = 0

chain = 5

v(0, 0) = 0
v(1, 0) = 0
v(2, 0) = 0
v(3, 0) = 1

v(0, 1) = 0
v(1, 1) = 0
v(2, 1) = 0
v(3, 1) = 0

For k = 1 To chain
    For i = 0 To 3
        For j = 0 To 3
            v(i, 1) = v(i, 1) + (v(j, 0) * Jar(j + 1, i + 1))
        Next j
```

```
      Next i

      For i = 0 To 3
          v(i, 0) = v(i, 1)
          v(i, 1) = 0
      Next i

      A = v(0, 0)
      B = v(1, 0)
      C = v(2, 0)
      D = v(3, 0)

      MsgBox "Step(" & k & ")=[" & A & "|" & B & "|" & C & "|"
  & D & "]"
  Next k

  End Sub
```

```
Output:
Step(1) = [0|0|1|0]
Step(2) = [0|0.5|0|0.5]
Step(3) = [0.25|0|0.75|0]
Step(4) = [0.25|0.375|0|0.375]
Step(5) = [0.4375|0|0.5625|0]
```

Supporting algorithm 13: Prediction framework based on a 4 × 4 transition matrix. Known transition probability values are directly used from a transition matrix for highlighting the behavior of a Markov chain. A variable number of states and configurations can be tested. For instance, a diagram with three states can be molded on the algorithm. However, since the framework is made for a total of four states, the following modifications are required: all Jar(n, 4) elements are set to zero (Jar(1, 4) = 0; Jar(2, 4) = 0; Jar(3, 4) = 0; and Jar(4, 4) = 0). Furthermore, the molding of a diagram with two states on this framework involves two modifications: all Jar(n, 4) elements are set to zero and all Jar(n, 3) elements are set to zero. Any type of diagram configuration can be tested by following two rules: (1) the absence of an arrow is indicated by zero, and (2) any value greater than zero and less than or equal to one indicates an arrow.

The current example takes into consideration case L from Table 8.1. Case L has a diagram with four states (state "A", "B", "C", and "D"), one of which is an absorbing state, namely state "A". Above, the output shows the prediction made on five steps, starting from state "D" (initial state vector: v = [0, 0, 0, 1]). Since the transition probabilities are provided by the state diagrams, the data can be entered manually inside the matrix. Returning to the jars and balls association, the format of the transition matrix (Jar) is: Jar(line, column) = probability

Figure 8.1 The state diagram of case A.

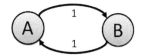

value, where the lines represent jars (jar "A" = 1, jar "B" = 2, jar "C" = 3, and jar "D" = 4) and the columns represent the types of balls (orange balls "A" = 1, blue balls "B" = 2, gray balls "C" = 3, and burgundy balls "D" = 4). For instance, the proportion of burgundy balls "D" inside jar "C", can be declared as: Jar(3, 4) = 1. Also, the proportion of orange balls "A" inside jar "B", can be written as: Jar(2, 1) = 0.5 (please see the state diagram and the transition matrix of case L). Thus, the above implementation can be used for a large number of experiments on Markov chains. Nevertheless, in this section, the above implementation has been used to predict the long-term (50 steps) behavior of each of the examples shown in Table 8.1.

8.2 Examples of Two-State Diagrams

8.2.1 Case A

Two states are considered: state "A" represented by jar "A" and state "B" represented by jar "B" (Figure 8.1). If only an arrow is emerging from a state and enters the other state, then the probability carried by each arrow will be 1 (Table 8.1A). The color of the ball from the current observation dictates the color of the jar from which the next draw is made. Thus, in order to satisfy the system from the diagram, in the current example, jar "A" must contain only balls "B" and jar "B" must contain only balls "A". Now the question would be how the system will behave once these draws are made? If draws start from state "A", then the sequence of draws will be periodical, such as: "ABABABABAB…"

However, if draws start from state "B", then the sequence of draws will appear asynchronous compared to the first: "BABABABABA…" This case *does not* represent a randomized process (as in throwing of a coin). Thus, there is no need for some special method to see that the average time spent in each state approaches $1/2$.

8.2.2 Case B

The state diagram of case A is further modified. A third arrow is introduced from state "B" to state "B". With this simple addition, the behavior of the system changes into a true randomized process (Table 8.1B and Figure 8.2). To calculate the probability carried by each arrow coming out of a state ($P[\text{from}|\text{to}]$),

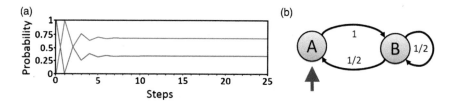

Figure 8.2 The state diagram of case B. (a) The behavior of the system on 25 steps. The orange line represents the probability that the system will be in state "A" while the blue line represents the probability that the system will be in state "B". (b) The state diagram of case B. The red arrow indicates the state (the jar) from which the first draw begins.

unity is divided by the number of arrows emerging from that state:

$$P[A|anywhere] = \frac{1}{\text{number of arrows that come out from state A}} = \frac{1}{1} = 1$$

$$P[B|anywhere] = \frac{1}{\text{number of arrows that come out from state B}} = \frac{1}{2} = 0.5$$

Thus, this implies that jar "A" contains only balls "B" ($P[A|anywhere] = 1/1 = 1$), and jar "B" contains $1/2$ balls "A" and $1/2$ balls "B" ($P[B|anywhere] = 1/2 = 0.5$).

The transition matrix m can be filled with probability values based on the data from the state diagram:

$$m = \begin{bmatrix} 0 & 1 \\ 0.5 & 0.5 \end{bmatrix}$$

Now the sequence of draws cannot be so easily predicted, but intuitively a sequence such as "ABBABBBABABABB..." can be expected, where "B" is more common than "A". Only if it exists, the stationary vector can indicate the average time spent in a state.

$$m = \begin{bmatrix} \alpha & (1-\beta) \\ (1-\alpha) & \beta \end{bmatrix} = \begin{bmatrix} 0 & 1-0.5 \\ 1-0 & 0.5 \end{bmatrix}$$

Since the proportion of balls is known directly from the state diagram, the average time spent in state "A" can be determined by:

$$\text{Time in A} = \frac{(1-\beta)}{(2-(\alpha+\beta))} = \frac{(1-0.5)}{(2-(0+0.5))} = 0.33 = 33\%$$

The average time spent in state "B" can be determined by:

$$\text{Time in B} = \frac{(1-\alpha)}{(2-(\alpha+\beta))} = \frac{(1-0)}{(2-(0+0.5))} = 0.66 = 66\%$$

Therefore, about 33% of the time draws will be made from jar "A" and 66% of the time draws will be made from jar "B".

8.3 Examples of Three-State Diagrams

8.3.1 Case C

Here, three states are shown in the simplest configuration: state "A" (represented by jar "A"), state "B" (represented by jar "B"), and state "C" (represented by jar "C"). The state diagram shows three arrows (Table 8.1C and Figure 8.3). As in the first case (case A), if only an arrow is emerging from a state and enters the other state, then the probability carried by each arrow will be 1 (Table 8.1C). In other words, in our example, probability 1 ($P[A|B] = P[B|C] = P[C|A] = 1$) means that jar "A" contains only balls "B", jar "B" contains only balls "C", and jar "C" contains only balls "A". Again, the periodicity of the system can be observed. If the system begins from state "A", then the sequence of draws will show a pattern such as: "ABCABCABC…" If the system begins from state "B", then the sequence of draws will be "BCABCABCA…", and if started from state "C", then the sequence of draws will be "CABCABCAB…" Regardless of the start state, in this configuration, the position of "A", "B", or "C" in the sequence of observations will present a periodicity. In conclusion, it is sufficient to consider any of the above possible sequences to find the average time spent in a state. For instance, consider the sequence "BCABCABCA…" which is composed of "BCA" pattern. The likelihood of "BCA" pattern in the sequence "BCABCABCA…" is $P[BCA] = 1$ and the probability of occurrence of each letter in this pattern is $P[B] = P[C] = P[A] = 1/3$. The average time spent in a state is the probability of each letter in the sequence "BCA". Thus, the average time spent on any of the jars is 33%.

8.3.2 Case D

The state diagram of case C is reused and modified as follows: two more arrows are introduced, one from state "B" to state "B", and one from state "A" to state

Figure 8.3 The state diagram of case C.

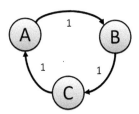

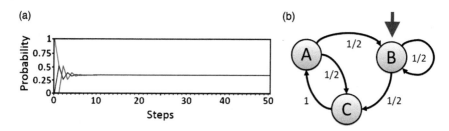

Figure 8.4 The state diagram of case D. (a) The behavior of the system on 50 steps. The orange line represents the probability that the system will be in state "A". The blue line represents the probability that the system will be in state "B". The gray line represents the probability that the system will be in state "C". (b) The state diagram of case D. The red arrow indicates the state (the jar) from which the first draw begins, in this case, state "B".

"C" (Table 8.1D and Figure 8.4). In this new case, the number of arrows in the diagram rose to five. To calculate the probability carried by each arrow coming out of a state, unity is divided by the number of arrows emerging from that state:

$$P[A|\text{anywhere}] = \frac{1}{\text{number of arrows that come out from state A}} = \frac{1}{2} = 0.5$$

$$P[B|\text{anywhere}] = \frac{1}{\text{number of arrows that come out from state B}} = \frac{1}{2} = 0.5$$

$$P[C|\text{anywhere}] = \frac{1}{\text{number of arrows that come out from state C}} = \frac{1}{1} = 1$$

Note that on the diagram, two arrows are coming out of state "A" (Table 8.1D). One goes to state "B" and one to state "C". Since the color of the ball from the current observation dictates the color of the jar from which the next draw is made, this implies that jar "A" must contain $\frac{1}{2}$ balls "B" and $\frac{1}{2}$ balls "C". Also, two arrows are coming out of state "B".

One goes back to state "B" and one goes to state "C". Thus, jar "B" also has an equal proportion of balls, namely $\frac{1}{2}$ balls "B" and $\frac{1}{2}$ balls "C". On the other hand, only one arrow is coming out of state "C" and points at state "A". This last observation implies that jar "C" must contain only balls "A". Next, curiosity drives us in calculating the average time spent in each state. In the present case, there is a total of three states (state "A", "B", and "C"), therefore $q = 3$. The transition matrix (P) derived from the state diagram can be used for obtaining the second matrix (m), needed for moving forward in solving the problem:

$$m_{ij} = \sum_{k=1}^{q} P_{i,k} \times P_{k,j}$$

In other words, $m = P \times P$, which can be written as:

$$m = P \times P = \begin{pmatrix} 0 & 0.5 & 0.5 \\ 0 & 0.5 & 0.5 \\ 1 & 0 & 0 \end{pmatrix} \times \begin{pmatrix} 0 & 0.5 & 0.5 \\ 0 & 0.5 & 0.5 \\ 1 & 0 & 0 \end{pmatrix} = \begin{pmatrix} 0.5 & 0.25 & 0.25 \\ 0.5 & 0.25 & 0.25 \\ 0 & 0.5 & 0.5 \end{pmatrix}$$

In order to find the average time ($t(s)$) spent in each state (or jar), the values from the matrix (m) columns are required for the next and final step. Thus, for the average time spent in state "A", the first column is used ($s = 1$):

$$t(1) = \left(\sum_{i=1}^{3} m_{i1} \right) \times \frac{100}{3} = (0.5 + 0.5 + 0) \times \frac{100}{3} = 33\%$$

For the average time spent in state "B", the second column is used ($s = 2$):

$$t(2) = \left(\sum_{i=1}^{3} m_{i2} \right) \times \frac{100}{3} = (0.25 + 0.25 + 0.5) \times \frac{100}{3} = 33\%$$

For the average time spent in state "C", the third column is used ($s = 3$):

$$t(3) = \left(\sum_{i=1}^{3} m_{i3} \right) \times \frac{100}{3} = (0.25 + 0.25 + 0.5) \times \frac{100}{3} = 33\%$$

Thus, a counterintuitive result is obtained for this diagram configuration. From a number of X steps, about 33% of the time draws will be made from jar "A", 33% of the time draws will be made from jar "B", and 33% of the time draws will be made from jar "C". The main conclusion is that case C and case D show the same average time. Nevertheless, the difference is that case D shows a steady-state vector while case C has a cyclical behavior and *does not* show a steady-state vector.

8.4 Examples of Four-State Diagrams

8.4.1 Case E

Consider four states, state "A" (represented by jar "A"), state "B" (represented by jar "B"), state "C" (represented by jar "C"), and state "D" (represented by jar "D"). For this case and the remaining cases, the rules are the same and further explanations may be redundant (Tables 8.1E–M). Thus, the transition probabilities between states can be found by simply looking at the diagram (Figure 8.5b). The probability carried by each arrow coming out of a state ($P[\text{from}|\text{to}]$) is the unity divided by the number of arrows emerging from that state:

$$P[\text{state}| \rightarrow] = \frac{1}{\text{Out}[\text{state}]}$$

(a)   (b)

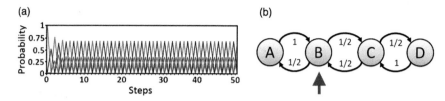

Figure 8.5 The state diagram of case E. (a) The behavior of the system on 50 steps. The orange line represents the probability that the system will be in state "A". The blue line represents the probability that the system will be in state "B". The gray line represents the probability that the system will be in state "C". The burgundy line represents the probability that the system will be in state "D". (b) The state diagram of case E. The red arrow indicates the state (the jar) from which the first draw begins, in this case, state "B".

where "state" can be state "A", "B", "C", or state "D". The arrow sign "→" represents any arrow leaving from that state to any other state. Thus, the above formula indicates that jar "B" contains $1/2$ orange balls of type "A" and $1/2$ gray balls of type "C" (Figure 8.5).

Also, jar "C" contains $1/2$ blue balls of type "B" and $1/2$ burgundy balls of type "D". Jar "A" contains only blue balls "B", and jar "D" contains only gray balls of type "C". Following the state diagram probabilities, a transition matrix (P) can be written as:

$$P = \begin{pmatrix} 0.0 & 1.0 & 0.0 & 0.0 \\ 0.5 & 0.0 & 0.5 & 0.0 \\ 0.0 & 0.5 & 0.0 & 0.5 \\ 0.0 & 0.0 & 1.0 & 0.0 \end{pmatrix}$$

What remains to be found out from this state diagram example is the average time ($t(s)$) spent in each state (or jar). First, the transition matrix (P) is multiplied by itself. As in the previous example, the result of this multiplication is the matrix (m):

$$m = P \times P = \begin{pmatrix} 0.50 & 0.00 & 0.50 & 0.00 \\ 0.00 & 0.75 & 0.00 & 0.25 \\ 0.25 & 0.00 & 0.75 & 0.00 \\ 0.00 & 0.50 & 0.00 & 0.50 \end{pmatrix}$$

Since matrix (m) has been obtained, the next step can begin. The formula for the average time spent in a state has previously been discussed in detail (Chapter 7):

$$t(s) = \left(\sum_{i=1}^{q} m_{is} \right) \times \frac{100}{q}$$

where q represents the total number of states and s represents one of the states. The present case is comprised of four states: state "A", "B", "C", and "D". Therefore the total number of states q will be equal to four ($q = 4$). Thus, for the average time spent in state "A", the first column is used ($s = 1$):

$$t(1) = \left(\sum_{i=1}^{4} m_{i1} \right) \times \frac{100}{4} = (0.5 + 0 + 0.25 + 0) \times \frac{100}{4} = 19\%$$

For the average time spent in state "B", the second column is used ($s = 2$):

$$t(2) = \left(\sum_{i=1}^{4} m_{i2} \right) \times \frac{100}{4} = (0 + 0.75 + 0 + 0.5) \times \frac{100}{4} = 31\%$$

For the average time spent in state "C", the third column is used ($s = 3$):

$$t(3) = \left(\sum_{i=1}^{4} m_{i3} \right) \times \frac{100}{4} = (0.5 + 0 + 0.75 + 0) \times \frac{100}{4} = 31\%$$

For the average time spent in state "D", the fourth column is used ($s = 4$):

$$t(4) = \left(\sum_{i=1}^{4} m_{i4} \right) \times \frac{100}{4} = (0 + 0.25 + 0 + 0.5) \times \frac{100}{4} = 19\%$$

The above method suggests that from a number of X steps, about 19% of the time draws will be made from jar "A", 31% of the time draws will be made from jar "B", 31% of the time draws will be made from jar "C", and about 19% of the time draws will be made from jar "D".

8.4.2   Case F

Here, four states are also used, however, in a different configuration. Notice that in this configuration, jar "A" contains $1/2$ blue balls of type "B" and $1/2$ gray balls of type "C" (Figure 8.6). Similarly, jar "D" contains $1/2$ blue balls of type "B" and $1/2$ gray balls of type "C". Also, jar "B" contains only orange balls "A", and jar "D" contains only gray balls of type "C".

A close observation of case F diagram indicates that $P[B|A] = 1$ and $P[C|D] = 1$ (Figure 8.6b). The two transition probabilities imply that letter "B" will always be followed by letter "A" and letter "C" will always be followed by letter "D". Also, since $P[D|B] = 1/2$ and $P[A|C] = 1/2$, the occurrence of "BA" or "CD" pair of letters/outcomes in a sequence of draws is $1/2$. Thus, if the system is started from jar "A", a sequence such as "ABACDBABACDCDBACD …" can be expected, where "BA" and "CD" pairs have the same probability of occurrence in the sequence. The two motif sequences ("BA" and "CD") with the same probability of occurrence can be analyzed together. The average time spent in a state can

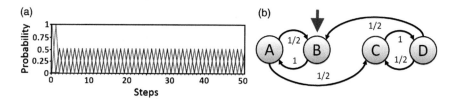

Figure 8.6 The state diagram of case F. (a) The behavior of the system on 50 steps. The orange line represents the probability that the system will be in state "A". The blue line represents the probability that the system will be in state "B". The gray line represents the probability that the system will be in state "C". The burgundy line represents the probability that the system will be in state "D". (b) The state diagram of case F. The red arrow indicates the state (the jar) from which the first draw begins.

be deduced from the probability of each letter in "BACD":

$$P[A] = 1/4 = 0.25$$
$$P[B] = 1/4 = 0.25$$
$$P[C] = 1/4 = 0.25$$
$$P[D] = 1/4 = 0.25$$

The time spent on average in each state can be expressed as $t = 100 \times P$. Note that the average time spent on each jar is about 25%. Above, the average time spent in each state has been determined through a step-by-step reasoning. In addition, the above values can be verified using the formula for the average time spent in a state:

$$t(s) = \left(\sum_{i=1}^{q} m_{is} \right) \times \frac{100}{q}$$

where q represents the total number of states and s represents one of the states. The transition matrix (P) derived from the state diagram can be written as:

$$P = \begin{pmatrix} 0.0 & 0.5 & 0.5 & 0.0 \\ 1.0 & 0.0 & 0.0 & 0.0 \\ 0.0 & 0.0 & 0.0 & 1.0 \\ 0.0 & 0.5 & 0.5 & 0.0 \end{pmatrix}$$

Next, the transition matrix (P) is multiplied by itself:

$$m = P \times P = \begin{pmatrix} 0.5 & 0.0 & 0.0 & 0.5 \\ 0.0 & 0.5 & 0.5 & 0.0 \\ 0.0 & 0.5 & 0.5 & 0.0 \\ 0.5 & 0.0 & 0.0 & 0.5 \end{pmatrix}$$

As in the previous example, the result of this multiplication is matrix (m) which is used to determine the average time spent in a state ($t(s)$). For the average time spent in state "A", the first column of the matrix is used ($s = 1$):

$$t(1) = \left(\sum_{i=1}^{4} m_{i1} \right) \times \frac{100}{4} = (0.5 + 0 + 0 + 0.5) \times \frac{100}{4} = 25\%$$

For the average time spent in state "B", the second column is used ($s = 2$):

$$t(2) = \left(\sum_{i=1}^{4} m_{i2} \right) \times \frac{100}{4} = (0 + 0.5 + 0.5 + 0) \times \frac{100}{4} = 25\%$$

For the average time spent in state "C", the third column is used ($s = 3$):

$$t(3) = \left(\sum_{i=1}^{4} m_{i3} \right) \times \frac{100}{4} = (0 + 0.5 + 0.5 + 0) \times \frac{100}{4} = 25\%$$

For the average time spent in state "D", the fourth column is used ($s = 4$):

$$t(4) = \left(\sum_{i=1}^{4} m_{i4} \right) \times \frac{100}{4} = (0.5 + 0 + 0 + 0.5) \times \frac{100}{4} = 25\%$$

The above method suggests that about 25% of the time draws will be made from jar "A", 25% of the time draws will be made from jar "B", 25% of the time draws will be made from jar "C", and about 25% of the time draws will be made from jar "D".

8.4.3 Case G

In this case, the probability of leaving state "B" through one of the arrows is 1/3 ($P[B|\rightarrow] = 1/3 = 0.33$), whereas the probability of leaving state "A", "C", or "D" is equal to 1 ($P[A|\rightarrow] = P[C|\rightarrow] = P[D|\rightarrow] = 1/1 = 1$). Thus, in the context of states as jars, notice that jar "A" and jar "C" contain only blue balls of type "B" and jar "D" contains only gray balls of type "C" (Figure 8.7). On the other hand, jar "B" contains 1/3 orange balls of type "A", 1/3 gray balls of type "C", and 1/3 burgundy balls of type "D".

The transition matrix (P) of the state diagram is used to make further determinations, such as the time spent on each state or the proportion of balls in the system. Thus, the corresponding transition matrix (P) of the state diagram is written as:

$$P = \begin{pmatrix} 0.00 & 1.00 & 0.00 & 0.00 \\ 0.33 & 0.00 & 0.33 & 0.33 \\ 0.00 & 1.00 & 0.00 & 0.00 \\ 0.00 & 0.00 & 1.00 & 0.00 \end{pmatrix}$$

(a)   (b)

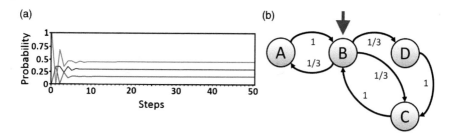

Figure 8.7 The state diagram of case G. (a) The behavior of the system on 50 steps. The orange line represents the probability that the system will be in state "A". The blue line represents the probability that the system will be in state "B". The gray line represents the probability that the system will be in state "C". The burgundy line represents the probability that the system will be in state "D". (b) The state diagram of case G. The red arrow indicates the state from which the first draw begins.

The transition matrix (P) is multiplied by itself:

$$m = P \times P = \begin{pmatrix} 0.33 & 0.00 & 0.33 & 0.33 \\ 0.00 & 0.66 & 0.33 & 0.00 \\ 0.33 & 0.00 & 0.33 & 0.33 \\ 0.00 & 1.00 & 0.00 & 0.00 \end{pmatrix}$$

The resulting transition matrix (m) from above is further used for determining the average time ($t(s)$) spent in each state (or jar). Just as in the other cases, for the average time spent in state "A", the first column is used ($s = 1$):

$$t(1) = \left(\sum_{i=1}^{4} m_{i1} \right) \times \frac{100}{4} = (0.33 + 0 + 0.33 + 0) \times \frac{100}{4} = 17\%$$

For the average time spent in state "B", the second column is used ($s = 2$):

$$t(2) = \left(\sum_{i=1}^{4} m_{i2} \right) \times \frac{100}{4} = (0 + 0.66 + 0 + 1) \times \frac{100}{4} = 41\%$$

For the average time spent in state "C", the third column is used ($s = 3$):

$$t(3) = \left(\sum_{i=1}^{4} m_{i3} \right) \times \frac{100}{4} = (0.33 + 0.33 + 0.33 + 0) \times \frac{100}{4} = 25\%$$

For the average time spent in state "D", the fourth column is used ($s = 4$):

$$t(4) = \left(\sum_{i=1}^{4} m_{i4} \right) \times \frac{100}{4} = (0.33 + 0 + 0.33 + 0) \times \frac{100}{4} = 17\%$$

Thus, in a series of draws, about 17% of the time draws will be made from jar "A", approximately 41% of the time draws will be made from jar "B", 25% of the time draws will be made from jar "C", and about 17% of the time draws will be made from jar "D".

8.5 Examples of State Diagrams Divided into Classes

The emergence of classes in the state diagrams of cases H, I, J, K, L, and M allows for interesting discussions from different angles. One of the topics concerns small repetitive units (patterns of letters) found in the sequence of observations. These repetitive units are generically named "motifs" and are of great importance in fields such as genetics and bioinformatics. Moreover, for these cases, the average time spent in a state can be directly analyzed without using the formula from above.

8.5.1 Case H

Case H shows more complex patterns in the sequence of draws. Here the key state of the system is represented by state "C" (Figure 8.8). In the context of states as jars, notice that jar "D" is filled with balls type "A", jar "A" is filled with balls type "B", and jar "B" is filled with balls type "C". The only jar of this system that has two types of balls is jar "C". Jar "C" contains $1/2$ balls type "B" and $1/2$ balls type "D". In case of a draw, from state "C", there are two possible versions of events: (1) from state "C" the system can switch to state "B" from which it returns directly to state "C" or (2) from state "C" the system makes the transition to state "D", from where it goes back to state "C" through states "A" and "B".

Thus, if draws are started from state "A", the system (diagram H) will produce sequences such as "A**BCBCBC**D**ABCBCBC**D**ABCDABC**". Since the only

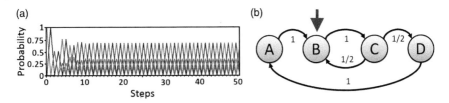

(a)   (b)

Figure 8.8 **The state diagram of case H.** (a) The behavior of the system on 50 steps. The orange line represents the probability that the system will be in state "A". The blue line represents the probability that the system will be in state "B". The gray line represents the probability that the system will be in state "C". The burgundy line represents the probability that the system will be in state "D". (b) The state diagram of case H. The red arrow indicates the state (the jar) from which the first draw begins.

node in the diagram is represented by state "C", the motif (patterns) sequences "BC" and "DABC" will have the same probability of occurrence in a sequence of draws, namely $P[BC] = P[DABC] = 1/2$. Two motif sequences ("BC" and "DABC") with the same probability of occurrence can be analyzed together. The time spent in a state on average can be deduced from the probability of each letter in "BCDABC":

$P[A] = 1/6 = 0.16$
$P[B] = 2/6 = 0.33$
$P[C] = 2/6 = 0.33$
$P[D] = 1/6 = 0.16$

Then, the time spent on average in each state can be expressed as $t = 100 \times P$. About 16% of draws will be made from jar "A" and 16% of draws will be made from jar "D". Notice that most often, draws will be made from jars "B" and "C" in an equal proportion. Thus, about 33% of draws will be made from jar "B" and 33% of draws will be made from jar "C".

8.5.2 Case I

Consider four states: state "A" (represented by jar "A"), state "B" (represented by jar "B"), state "C" (represented by jar "C"), and state "D" (represented by jar "D"). Once again each state is considered as a unity, namely 1. The number of arrows that depart from a state divides this unity. In this particular case, three arrows depart from state "A", then the probability of leaving state "A" through one of the arrows is the number of arrows that divide the unity, namely 1/3 ($P[A|\text{anywhere}] = 1/3 = 0.33$). One arrow departs from state "B", then the probability of leaving state "B" through that arrow is 1/1 ($P[B|\text{anywhere}] = 1/1 = 1$). Also, only one arrow departs from state "D", thus the probability of leaving state "D" through that arrow is 1/1 ($P[D|\text{anywhere}] = 1/1 = 1$). The same is true for state "C", therefore $P[C|\text{anywhere}] = 1/1 = 1$ (Table 8.1E). A series of observations are now apparent from these probability values, namely that jar "A" contains 33% balls "B", 33% balls "C" and 33% balls "D". Also, jar "B" contains only balls "A", jar "C" contains only balls "D", and jar "D" contains only balls type "C". In the examples above, so far all states were reachable. Here, in this example, state "A" receives no arrows, but sends one arrow to state "D", one arrow to state "C", and one arrow to state "B" (Figure 8.9). Since state "A" receives no arrows, it is not reachable from other states and is not a part of the system. State "B" receives an arrow from state "A" and sends one arrow to state "D". Since state "A" receives no arrows, state "B" is not reachable from other states and is not a part of the system. Now the question would be how the system will behave once these draws are started in this configuration? If the system is started from state "A" (jar "A"), then the sequence of draws will be: "ACDCDCDCD…" or "ADCDCDCDC…" or "ABDCDCDCD…", with an equal probability of occurrence, namely 1/3. If the system starts from state "B",

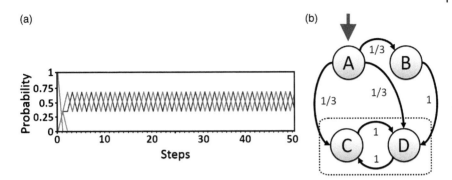

Figure 8.9 The state diagram of case I. (a) The behavior of the system on 50 steps. The orange line represents the probability that the system will be in state "A". The blue line represents the probability that the system will be in state "B". The gray line represents the probability that the system will be in state "C". The burgundy line represents the probability that the system will be in state "D". (b) The state diagram of case I. The red arrow indicates the state (the jar) from which the first draw begins, in this case, state "A".

then the sequence of observations will be: "BDCDCDCDC..." If state "D" is the initial state from which the system begins, then the sequence of draws will be periodical, such as: "DCDCDCDCD..." However, if the system begins from state "C", then the sequence will be asynchronous compared to the previous one: "CDCDCDCDC..." Thus, no matter from which state the system begins, as the number of draws goes to infinity, the proportion of "C" observations will approach $1/2$ and the proportion of "D" observations will approach $1/2$.

8.5.3 Case J

Case J shows an absorbing class that consists of states "C" and "D". States "A" and "B" are **transient** since the system eventually reaches "C" and "D" and does not return to "A" and "B" (Figure 8.10). On the other hand, states "C" and "D" are

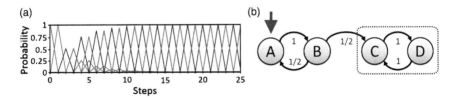

Figure 8.10 The state diagram of case J. (a) The behavior of the system on 25 steps. The orange line represents the probability that the system will be in state "A". The blue line represents the probability that the system will be in state "B". The gray line represents the probability that the system will be in state "C". The burgundy line represents the probability that the system will be in state "D". (b) The state diagram of case J. The red arrow indicates the state (the jar) from which the first draw begins, in this case, state "A".

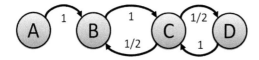

Figure 8.11 The state diagram of case K.

recurrent as the system goes from "C" to "D" and vice versa until the machine is stopped. Thus, the system will enter in the second class once the transition is made from "B" to "C". Given the above observations, case F could produce a sequence of observations like "ABABA**BCDCDCDCDCD**..." where "C" and "D" appear periodically in the sequence. Again, no matter from which state the system begins, as the number of draws goes to infinity, the proportion of "C" observations will approach $\frac{1}{2}$ and the proportion of "D" observations will approach $\frac{1}{2}$.

8.5.4 Case K

In case K, state "A" is **transient** since the system immediately reaches another separate class that consists of **recurrent** states "B", "C", and "D", and does not return to "A" (Figure 8.11). Since the transitions from "B" to "C" and from "D" to "C" have a probability value of 1 ($P[B|C] = 1$ and $P[D|C] = 1$), "C" will appear periodically in sequence. For instance, if the system is started from state "A", the expectation consists of a sequence of observations such as "ABCXCXCXCXCX...", where "X" represents state "B" or state "C". In the context of infinity, the first two observations ("AB...") can be irrelevant. Regardless of the state from which the system begins, as the number of draws goes to infinity, the proportion of "C" observations will approach $\frac{1}{2}$ ($P[C] = \frac{1}{2}$). Also, as the number of draws goes to infinity, the proportion of "X" observations will approach $\frac{1}{2}$ ($P[X] = \frac{1}{2}$). Since the transition from state "C" to state "B" and the transition from state "C" to state "D" have equal chances, "X" will be replaced 50% of the time by "B" and 50% of the time by "D" ($P[C|B] = P[C|D] = \frac{1}{2}$). Also notice that:

$$P[C] + P[X] = \left(\frac{1}{2}\right) + \left(\frac{1}{2}\right) = 1$$

Thus, the probability of "X" in the sequence of observations will be $P[X] = 1 - P[C] = \frac{1}{2}$. Since the value of $P[X]$ is $\frac{1}{2}$, the $P[B]$ and $P[D]$ values can also be found:

$$P[B] = \frac{P[X]}{P[C|B]} = \frac{1/2}{1/2} = 0.25$$

$$P[D] = \frac{P[X]}{P[C|D]} = \frac{1/2}{1/2} = 0.25$$

Also notice that:

$$P[C] + P[B] + P[D] = 0.5 + 0.25 + 0.25 = 1$$

Then, the time spent on average in each of the three states can be expressed as $t = 100 \times P$. About 25% of draws will be made from jar "B" and 25% of draws will be made from jar "D", whereas roughly 50% of draws will be made from jar "C".

8.6 Examples of State Diagrams with Absorbing States

Note that the values on the main diagonal of the transition matrices represent the probability of moving from a state to the same state (Tables 8.1F–M). The system contains an *absorbing state* if any of the elements on the main diagonal of the matrix are equal to 1. When the system eventually jumps to an *absorbing state*, it will remain in that state for the rest of the steps taken since the probability of moving back is always a certainty. The only cases in which a value of 1 can be observed on the main diagonal of the matrices are cases L and M from Table 8.1.

8.6.1 Case L

Again, consider four states: state "A" (represented by jar "A"), state "B" (represented by jar "B"), state "C" (represented by jar "C"), and state "D" (represented by jar "D"). Same as in other cases, the probability carried by each arrow coming out of a state must be found. Thus, unity is divided by the number of arrows emerging from that state:

$$P[\text{state}|\rightarrow] = \frac{1}{\text{Out}[\text{state}]}$$

where "state" can be state "A", "B", "C", or state "D". The arrow sign "→" represents any arrow leaving from that state to any other state. Thus, by interpreting the results of the above formula, the following observations can be made: Jar "B" contains $1/2$ orange balls of type "A" and $1/2$ gray balls of type "C". Jar "C" contains $1/2$ blue balls of type "B" and $1/2$ burgundy balls of type "D". Also, jar "A" contains only orange balls of type "A" and jar "D" contains only gray balls of type "C" (Figure 8.12). These observations are important for an overview of the system. Notice that case L has only one absorbing state, namely state "A" (Figure 8.12). It is reasonable to speculate that if this system is started from state "D", the expected sequence of observations may look like: "DCDCBC**BA**AAA…" Also, if the system is started from state "C", the expected sequence of observations may look like: "CDCBC**BA**AAAA…" In case L, the main interest lies in finding the probability of reaching the absorbing state (jar "A") after a certain number of steps (extraction of balls). The convergence toward the steady-state vector represents the natural path of the machine to equilibrium. An interesting strategy for understanding this four-state system is to observe how it will behave until it reaches the stationary vector. Since the system can be started from one of

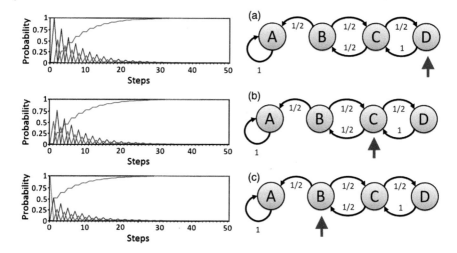

Figure 8.12 The state diagram and the convergence path of case L. (a) The behavior of the system on 50 steps when starting from state "A". (b) The behavior of the system on 50 steps when starting from state "B". (c) The behavior of the system on 50 steps when starting from state "C". In any of the three charts, the orange line represents the probability that the system will be in state "A". The blue line represents the probability that the system will be in state "B". The gray line represents the probability that the system will be in state "C", and the burgundy line represents the probability that the system will be in state "D". The red arrows indicate the state (the jar) from which the first draw begins.

the three states (namely state "D", state "C", or state "B"), three initial state vectors can be tested. In our example above, the prediction has been made for 50 steps. If the vector components are plotted on a graph for 50 steps, an overview for the long-run path (convergence toward the steady-state vector) of the system can be observed (Figure 8.12). The probability that the system will be in state "A" grows after every two steps, whereas the probability that the system will be in state "C", "B", or "D" decreases synchronously after every two steps (Figure 8.12). Also, all three charts show that in the range of just 50 steps, the system reaches state "A" regardless of the initial state vector (Figures 8.12a–c).

8.7 The Gambler's Ruin

8.7.1 Case M

Case M has two absorbing states, namely state "A" and state "D". If the system is started from state "B", two outcomes can be expected: (1) a sequence of observations such as "BCB**CD**DDDDD…" or (2) a sequence of observations such as "BCBC**BA**AAAAA…" A classic example of a Markov chain with two

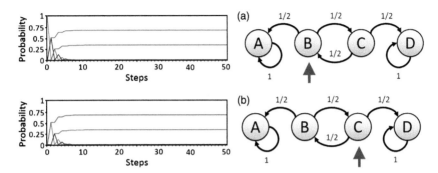

Figure 8.13 The state diagram and the convergence path for case M. (a) The behavior of the system on 50 steps when starting from state "A". (b) The behavior of the system on 50 steps when starting from state "B". The orange line represents the probability that the system will be in state "A". The blue line represents the probability that the system will be in state "B". The gray line represents the probability that the system will be in state "C", and the burgundy line represents the probability that the system will be in state "D". The red arrows indicate the state (the jar) from which the first draw begins.

absorbing states is the "Gambler's Ruin". A game with fixed odds involving the same bet is representative of the current example (Figure 8.13). An individual enters a casino with 50 euros. He decides to play roulette. At each spin of the roulette, he is betting 25 euros on red. If the ball falls on red, he wins 25 euros and if the ball falls on black, he loses 25 euros. Since red and black on a roulette are in equal proportions, he has a 50% chance of winning. The player will quit the game in two situations: (1) remains without money and (2) wins 75 euros (with 25 euros more than the initial amount by which he entered the casino). The modeling of this process can be done with a transition diagram with four states. State "A" represents 0 euros, state "B" represents 25 euros, state "C" represents 50 euros, and state "D" represents 75 euros. If he remains without money, then he no longer has what to bet and remains in state "A". If he wins 75 euros, he will quit the game and remains in state "D". However, when he enters in the casino with 50 euros, he starts directly in state "C". Since he starts from state "C" and makes the first bet, he has $1/2$ chance to move into state "B" (loses 25 euros) or $1/2$ chance to move into state "D" (wins 25 euros). If he goes to state "D", then he won 25 euros which adds to the 50 euros which he already had and makes a total of 75 euros. If he goes to state "B", then he lost 25 euros from a total of 50 euros and remains with 25 euros. If he makes a new bet from state "B", he has $1/2$ chance to move into state "A" (loses 25 euros) or $1/2$ chance to move into state "C" (wins 25 euros). Thus, these are the possible events for this player. If he enters the game with 50 euros, then he starts in state "C". Thus, the initial state vector will be:

$$v = [P[\text{A}] \quad P[\text{B}] \quad P[\text{C}] \quad P[\text{D}]]$$

Since the probability of starting in state "C" is a certain event ($P[C] = 1$), the probability values of the other components of the initial state vector are equal to zero:

$$v = [0 \quad 0 \quad 1 \quad 0]$$

Let us consider that relevant information would be to know something about the success or failure of this player after 20 bets on red. What would be likely to happen after 20 bets on red, knowing that the player enters the game with 50 euros?

$$[0 \quad 0 \quad 1 \quad 0] \begin{pmatrix} 1.0 & 0.0 & 0.0 & 0.0 \\ 0.5 & 0.0 & 0.5 & 0.0 \\ 0.0 & 0.5 & 0.0 & 0.5 \\ 0.0 & 0.0 & 0.0 & 1.0 \end{pmatrix}^{20} = [0.333 \quad 0 \quad 0 \quad 0.666]$$

If he enters the game with 50 euros, then after 20 bets on red, the chance of winning 75 euros is 66% and the chance of losing all the money is 33%. Note that after 20 bets, the probability that the player will be in state "B" or state "C" is zero (Figure 8.13). Now what would be likely to happen knowing that the player enters the game with 25 euros? If he enters the game with 25 euros, then he starts in state "B":

$$[0 \quad 1 \quad 0 \quad 0] \begin{pmatrix} 1.0 & 0.0 & 0.0 & 0.0 \\ 0.5 & 0.0 & 0.5 & 0.0 \\ 0.0 & 0.5 & 0.0 & 0.5 \\ 0.0 & 0.0 & 0.0 & 1.0 \end{pmatrix}^{20} = [0.666 \quad 0 \quad 0 \quad 0.333]$$

If he enters the game with 25 euros, then after 20 bets, the chance of winning 75 euros is 33% and the chance of losing all the money is 66%. Regardless of the initial state vector, the probability that the player will be in state "B" or state "C" is zero if the number of bets tends to infinity.

9

The Simulation of an *n*-State Markov Chain

9.1 Introduction

A Markov chain represents the behavior of a random process which may only find itself in a number of different states. The process moves from a state to another in discrete times. One natural way to answer questions about Markov chains is to simulate them. An important phase when implementing a Markov chain consists in the ability to test the predictions by comparing them with the results from a simulation. In other words, instead of real jars and balls, an algorithm can be used to simulate the process. In Chapter 2, a simulation of a chain with two states has been made (two jars and two types of balls). However, in the previous chapter, different configurations of chains have been discussed. Furthermore, those configurations of chains have used a different number of states, from two states up to four states. What can be accomplished by a simulation of a Markov chain? One answer would be the validation of the expected average time (from prediction) by a simple comparison with the actual average time (from simulation). Whether a chain has a stationary vector or not, the time spent in each state converges toward an average. A second answer is that a chain simulation can mimic the behavior of some complex systems, allowing numerous applications for behavior prediction. Therefore, this chapter is dedicated to the simulation of *n*-state Markov chains in different configurations. Special attention has been paid in the simulation of several Markov diagrams from Chapter 8, namely: Case A, Case L, Case F, and Case G.

9.2 The Simulation of Behavior

A few steps are needed in order to move forward from a Markov diagram to the simulation process. In the first step, a familiar example of a Markov diagram is considered (similar to that used in Chapter 7) and a transition matrix is built (Figure 9.1). In the second phase, the Markov diagram is placed in the context of an experiment based on a set of jars and balls of different colors. In a third

Markov Chains: From Theory to Implementation and Experimentation, First Edition. Paul A. Gagniuc.
© 2017 John Wiley & Sons, Inc. Published 2017 by John Wiley & Sons, Inc.
Companion website: www.wiley.com/go/gagniuc/markovchains

phase, the jars and balls of different colors are translated in a new representation (electronic format) meant to be used by an algorithm for simulation purposes. In the conclusive phase, the electronic representation is used in the context of an algorithm that simulates draws.

9.2.1   The State Diagram Example

In general, theory defines the number of states in a diagram. For example, if the weather is taken into account, a model can consider two states: "no rain" (state "A") and "rain" (state "B"). Three states can also be considered: "no rain" (state "A"), "moderate rain" (state "B"), and "heavy rain" (state "C"). By the same rationale, other diagrams consisting of four states (i.e., states: "fog", "no rain", "moderate rain", and "heavy rain") up to n states can be constructed. Nevertheless, here the example is limited to three states: state "A", state "B", and state "C" (Figure 9.1a). Also, the transition probability values can be attached to the diagram depending on various external factors, such as transition probability values derived from a series of observations captured from a physical process. Nevertheless, for simplicity, on this ideal system (shown below) the transition probabilities are inferred relying on arrows seen in the diagram. Thus, three arrows leave state "A", two arrows leave state "B", and only one arrow leaves state "C". What can be said about this system? Consider each state as a unity, namely 1. The number of arrows that depart from a state divides this unity. For instance, if three arrows depart from state "A", then the probability of leaving state "A" through one of the arrows is the number of arrows that divide the unity, namely $1/3$ ($P[A|anywhere] = 1/3 = 0.33$). If two arrows depart from state "B", then the probability of leaving state "B" through one of the arrows is $1/2$ ($P[B|anywhere] = 1/2 = 0.5$). However, only one arrow departs from state "C", thus the probability of leaving state "C" through that arrow is $1/1$ ($P[C|anywhere] = 1/1 = 1$). Thus, a transition matrix can be constructed according to the diagram configuration (Figure 9.1b).

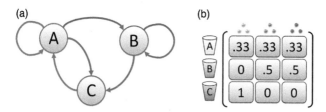

Figure 9.1 Example of state diagram. (a) The state diagram of a chain with three states. (b) The matrix lines represent the inside of the orange, blue, and red jars. The matrix columns represent the proportion of balls of different colors, namely orange balls, blue balls, and red balls.

9.2.2   From Markov Diagrams to Jars and Balls

Now imagine a system built from jars and balls. Consider that states "A", "B", and "C" are jars of different colors filled with balls of different colors (Figure 9.2). Previous examples have shown that probabilities stored by the transition matrix reflect the proportion of balls in each jar (Figure 9.1b). Assume that each row of the transition matrix represents one jar of a certain color: jar "A" (orange), jar "B" (blue), and jar "C" (red). Each column of the transition matrix represents the proportions of balls present in each jar. The first column represents the proportion of orange balls in each jar, the second column represents the proportion of blue balls in each jar, and the third column represents the proportion of red balls in each jar. The first row in the matrix shows that the orange jar "A" contains an equal proportion of balls of each color, namely 33% orange balls, 33% blue balls, and 33% red balls (Figure 9.2). The blue jar "B" contains 50% orange balls and 50% blue balls, and the red jar "C" contains only orange balls. Assume that a series of extractions are made respecting the chain rule in which the color of the ball indicates the color of the jar from which the next extraction shall be made. Suppose 10,000 extractions are performed, which of course, involve 10,000 observations. What would be the expected long-term

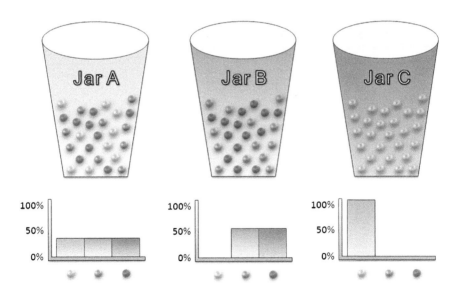

Figure 9.2 Translation of Markov diagrams to jars and balls. Each jar contains 27 balls of different colors. Each graph is showing the proportion of balls in the above jar. The orange Jar "A" contains an equal proportion of balls of each color (33%). The blue Jar "B" contains 50% blue balls and 50% red balls. The red Jar "C" contains in exclusivity orange balls.

observations? Of course, it would be a time-consuming experiment. This is the main reason for which such an experiment is made using a computer.

9.2.3 Representation of the Jars

The exact number of balls in each jar is unknown; however, the proportions of balls are known from the transition matrix. Since the proportions are known, there is a quite large degree of freedom for representing jars. Thus, a string of n letters may represent a jar with n balls, where n can be arbitrarily chosen. Such a string may be composed of "A", "B", or "C" letters that represent the proportions of orange, blue, and red balls in a jar. In this case, a series of 27 letters are used for each jar (27 × 3 jars = 81 letters system wide). The question that arises is: what is the number of "A", "B", and "C" letters if a total of 27 letters (L_{tot}) are used? For instance, the most obvious is the case of jar "C" which contains 100% balls "A". In electronic format, the content of the red jar "C" is represented by 27 letters "A" (Figure 9.2). Moving forward, the yellow jar "A" contains 33% balls "A", "B", and "C". Intuitively, nine letters represent 33% from a total of 27 letters. Thus, nine letters "A", nine letters "B", and nine letters "C" are used for jar "A" representation. These are cases that are easily determined by means of simple reasoning. However, a method must be used when proportions vary in a manner that does not allow for simplicity. The multiplication of probability values by the total number of letters (L_{tot}) provides a direct approach to actually calculate the number of letters. A letter is an indivisible entity that must be represented by integers. In order to obtain integers, the values provided by the following formulas have to be rounded. Thus, the number of "A", "B", and "C" letters inside the orange jar (Jar_A) is:

$$Jar_A[A] = L_{tot} \times P_A[A] = 27 \times 0.33 = 8.91 = {\sim}9$$
$$Jar_A[B] = L_{tot} \times P_A[B] = 27 \times 0.33 = 8.91 = {\sim}9$$
$$Jar_A[C] = L_{tot} \times P_A[C] = 27 \times 0.33 = 8.91 = {\sim}9$$

Note that the number of "A", "B", and "C" letters from the orange jar representation is equal to L_{tot}:

$$Jar_A[A] + Jar_A[B] + Jar_A[C] = L_{tot} = 27$$

By continuing this rationale, the proportion of "A", "B", and "C" letters from the blue jar (Jar_B) is:

$$Jar_B[A] = L_{tot} \times P_B[A] = 27 \times 0 = 0$$
$$Jar_B[B] = L_{tot} \times P_B[B] = 27 \times 0.5 = 13.5 = {\sim}13$$
$$Jar_B[C] = L_{tot} \times P_B[C] = 27 \times 0.5 = 13.5 = {\sim}14$$

We again note that the number of "A", "B", and "C" letters from the blue jar representation is equal to L_{tot}:

$$\text{Jar}_B[A] + \text{Jar}_B[B] + \text{Jar}_B[C] = L_{tot} = 27$$

However, an exception occurs in the case of the blue jar. One number is rounded down and the other number is rounded up (see Section 9.2.4). And last, the number of "A", "B", and "C" letters from the red jar (Jar$_C$) can be found:

$$\text{Jar}_C[A] = L_{tot} \times P_C[A] = 27 \times 1 = 27 = 27$$
$$\text{Jar}_C[B] = L_{tot} \times P_C[B] = 27 \times 0 = 0$$
$$\text{Jar}_C[C] = L_{tot} \times P_C[C] = 27 \times 0 = 0$$
$$\text{Jar}_C[A] + \text{Jar}_C[B] + \text{Jar}_C[C] = L_{tot} = 27$$

Up to this point, three distinct string representations have been used, namely: Jar$_A$, Jar$_B$, and Jar$_C$ which are the rough copies of the original jars:

$$\text{Jar}_A = (\text{AAAAAAAAABBBBBBBBBCCCCCCCCC})$$
$$\text{Jar}_B = (\text{BBBBBBBBBBBBBCCCCCCCCCCCCCC})$$
$$\text{Jar}_C = (\text{AAAAAAAAAAAAAAAAAAAAAAAAAAA})$$

Thus, up to this point real jars have been represented by using strings of letters. The representation of Jar$_A$ contains nine letters "A", nine letters "B", and nine letters "C". Also, Jar$_B$ contains 0 letters "A", 13 letters "B" and 14 letters "C". And last, the representation of Jar$_C$ contains 27 letters "A" and none of the other two.

9.2.4 Discussion on the Imbalance of Chances

Note that letters are represented by integers. In order to obtain integers, the values provided by the above expressions have to be rounded. For instance, in the case of the blue jar: $27 \times 0.5 = 13.5$ letters "B" or "C". The questions that arise are: How could this number be rounded up or down in order to provide an equal number of "B" and "C" letters in a set of 27? In favor of which state can a letter be added? One more "B" letter (and one "C" letter in minus) in the blue jar means that the chances of returning to the blue jar increase slightly (implicitly, the chances of moving to the red jar decrease slightly). On the other hand, one more "C" letter (and one "B" letter in minus) in the blue jar means that the likelihood of moving to the red jar increases slightly. Of course, the problem can be solved if all the values *are always rounded down*, which, results in a *different number of letters for each jar*. This means that 13.5 letters "B" become 13 letters "B" and 13.5 letters "C" also become 13 letters "C" (13 + 13 = 26 letters for jar "B", 27 letters for jar "A", and 27 letters for jar "C"). However, *the aim* is to keep a *constant number* of 27 letters in *each jar*. The only option is to randomly choose a letter in favor of one or the other state. Therefore, one number is rounded down and the other number is rounded up. Thus, assume that 13.5 letters "B" become 13 letters "B" (rounded down) and 13.5 letters "C" become 14 letters "C" (rounded up; 13 + 14 = 27 letters for jar "B"). For short strings of letters, probabilities are drastically changed in such situations in favor of one of the states. To avoid these cases, very long strings of letters may be used as a solution. If a letter is chosen in favor of one of the states, then the bias (that ±1 letter) remains negligible. In other words, the longer the string of letters representing a jar, the lower is the bias. Let us take two examples, a string of 11 letters which represents the proportions of balls in jar "B" and another string of 27 letters, which also represents the proportions of balls in jar "B" (Figures 9.3a–c). When the representation is made by using only 11 letters, the proportions of balls inside jar "B" are obviously impossible to be accurately replicated. In the best case scenario, 11 letters can be partitioned in 5 letters "B" that represents 45% instead of 50% ((100/11) × 5 = 45%), and 6 letters "C" representing 55% instead of 50% ((100/11) × 6 = 55%). When choosing a random letter from a

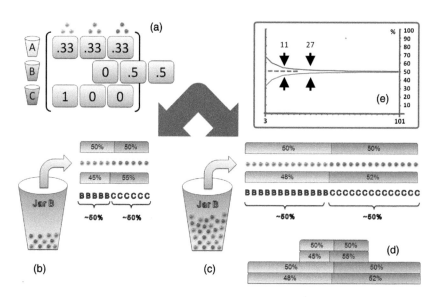

Figure 9.3 An imbalance of chances. (a) Shows the row of the transition matrix taken into discussion. The row stores the transition probabilities corresponding to jar "B" which translate into two representations: (b) first, by using a short sequence of 11 letters; (c) second, by using a long sequence of 27 letters. (d) Shows the difference of chance between a short sequence of 11 letters and a long sequence of 27 letters. (e) By using prime numbers, the distribution shows the convergence of the proportions of letters to the actual probability values indicated by the transition matrix. The length of the string (L_{tot}) is presented on the x-axis and the proportion of "B" (the blue line) and "C" letters (the red line) on the y-axis.

total of 11 letters, the probability of choosing letter "B" is 0.45 (instead of 0.5), and the probability of choosing letter "C" is 0.55 (instead of 0.5). But when the representation is made using 27 letters, the proportions tend closer to reality. As discussed earlier, in this case, the proportions may be represented by 13 letters "B" (representing 48%) and 14 letters "C" (representing 52%). Thus, when choosing a random letter from a total of 27 letters, the probability of choosing letter "B" is 0.48, and the probability of choosing letter "C" is 0.52. With a difference of only 16 letters between the two examples (27 − 11 = 16), one may observe *a convergence* of 3% (48 − 45 = 3%; 55 − 52 = 3%) toward the ideal proportions indicated in the transition matrix (Figure 9.3d). This means that with an additional 16 letters, the probability of choosing letter "B" increased from 0.45 to 0.48, and the probability of choosing letter "C" decreased from 0.55 to 0.52. At least in theory it can be said that as the total number of letters (L_{tot}) increases to infinity, the proportion of letters will approach 1/2 (50% for "B" and 50% for "C"). Thus, if a constant number of letters for each jar is desired for various experimental reasons, then the solution lies in the generation of very long string representations of jars to avoid an imbalance of chances inside the system. Such an imbalance of chances is typically encountered when the multiplication of probability values by the total number of letters (L_{tot}) does not result in an integer value. One way to observe this convergence consists in using prime numbers for a simple experiment (Figure 9.3e and Table 9.1). A prime number is a number that has no positive divisors other than 1 and itself. Consider the first 25 prime numbers, from 3 to 101. Also, consider 25 separate strings ($S_{1...25}$) representative of the same jar "B". If the total number of letters (L_{tot}) of each string ($S_{1...25}$) is equal to 3, or 5, or 7, or 13, ..., or 101, a convergence toward the ideal proportions indicated by the transition matrix can be observed (Table 9.1). For instance, consider the first prime number in the list, namely 3. Thus, 3 represents the total number of letters (L_{tot} = 3) in the first string (S_1). The result of the multiplication of probability values by the total number of letters (L_{tot}) is then used to determine the number of "B" letters ($Jar_B[B]$) and the number of "C" letters ($Jar_B[C]$) required for the simulation of the proportions (i.e., $P_B[B]$ and $P_B[C]$) indicated by the transition matrix:

$$Jar_B[A] = L_{tot} \times P_B[A] = 3 \times 0 = 0$$
$$Jar_B[B] = L_{tot} \times P_B[B] = 3 \times 0.5 = 1.5$$
$$Jar_B[C] = L_{tot} \times P_B[C] = 3 \times 0.5 = 1.5$$

In the case of the blue jar, an equal proportion of "B" and "C" letters should be used in order to satisfy the proportions indicated by the transition matrix. But at the same time, integers are needed for the representation of the letters. As before, in order to obtain integers one number is rounded down ($Jar_B[B]$ = 1.5 = 1) and the other number is rounded up ($Jar_B[C]$ = 1.5 = 2). Thus, two

Table 9.1 Convergence to imposed ratios. Prime numbers are used for testing the convergence of the proportions of letters to the actual probability values indicated by the transition matrix.

| No. | Representative strings of jar "B" | | L_{tot} (prime) | $L_{tot} \times$ 0.5 | The proportion of "B" letters (B%) | The proportion of "C" letters (C%) | Convergence |
	Number of "B" letters (rounded down)	Number of "C" letters (rounded up)					
S_1	1	2	3	1.5	33.33	66.67	16.67
S_2	2	3	5	2.5	40.00	60.00	10.00
S_3	3	4	7	3.5	42.86	57.14	7.14
S_4	5	6	11	5.5	45.45	54.55	4.55
S_5	6	7	13	6.5	46.15	53.85	3.85
S_6	8	9	17	8.5	47.06	52.94	2.94
S_7	9	10	19	9.5	47.37	52.63	2.63
S_8	11	12	23	11.5	47.83	52.17	2.17
S_9	14	15	29	14.5	48.28	51.72	1.72
S_{10}	15	16	31	15.5	48.39	51.61	1.61
S_{11}	18	19	37	18.5	48.65	51.35	1.35
S_{12}	20	21	41	20.5	48.78	51.22	1.22
S_{13}	21	22	43	21.5	48.84	51.16	1.16
S_{14}	23	24	47	23.5	48.94	51.06	1.06
S_{15}	26	27	53	26.5	49.06	50.94	0.94
S_{16}	29	30	59	29.5	49.15	50.85	0.85
S_{17}	30	31	61	30.5	49.18	50.82	0.82
S_{18}	33	34	67	33.5	49.25	50.75	0.75
S_{19}	35	36	71	35.5	49.30	50.70	0.70
S_{20}	36	37	73	36.5	49.32	50.68	0.68
S_{21}	39	40	79	39.5	49.37	50.63	0.63
S_{22}	41	42	83	41.5	49.40	50.60	0.60
S_{23}	44	45	89	44.5	49.44	50.56	0.56
S_{24}	48	49	97	48.5	49.48	50.52	0.52
S_{25}	50	51	101	50.5	49.50	50.50	0.50

integers have been obtained, but the proportions indicated by the transition matrix were only partially represented:

$$B\% = \left(\frac{100}{L_{\text{tot}}}\right) \times \text{Jar}_B[B] = \left(\frac{100}{3}\right) \times 1 = 33\%$$

$$C\% = \left(\frac{100}{L_{\text{tot}}}\right) \times \text{Jar}_B[C] = \left(\frac{100}{3}\right) \times 2 = 66\%$$

Nevertheless, by using the same reasoning for the remaining cases ($S_{2\ldots25}$), a convergence of the two proportions of letters ("B" and "C") can be observed (Figure 9.3e and Table 9.1). For instance, in the case of S_{25} ($L_{\text{tot}} = 101$), when choosing a random letter from a total of 101 letters, the probability of choosing letter "B" increases to 0.494, and the probability of choosing letter "C" decreases to 0.506. The longer the sequence of letters (L_{tot}), the greater the convergence to the real proportion indicated by the transition matrix. At least in theory it can be said that as the total number of letters (L_{tot}) increases to infinity, the proportion of "B" and "C" letters will approach 1/2 (50% for "B" and 50% for "C"). Again, under normal circumstances, an uneven number of letters (L_{tot}) for each jar eliminates the problem of a chance imbalance.

9.2.5 Simulation of the System

In order to perform a computer simulation, all jars (Jar_A, Jar_B, and Jar_C) can be represented by one vector (Jar) with three components: Jar(1), Jar(2), and Jar(3). Note that jar "A" is associated with index "1", jar "B" corresponds to index "2", and jar "C" is identified by index "3" (Figure 9.4). The index allows the accession of the vector components from the perspective of an algorithm implementation. Each component contains a sequence of 27 letters that represents the proportions of balls from a jar. Therefore, consider that all the balls from the orange jar (Jar_A) are represented by the first component, namely Jar(1); all the balls from the blue jar (Jar_B) are represented by the second component Jar(2); all the balls from the red jar (Jar_C) are represented by the third component Jar(3). In order to simulate a draw, a random integer is generated between 1 and 27. This random integer represents the letter found at ith position on a string of 27 letters. Thus, whether a ball is randomly chosen from a real jar containing 27 balls or a letter is randomly selected from a string of 27 letters, the process is exactly the same. Nevertheless, how can draws be simulated on a computer? Suppose a number of 20 draws are simulated from this three-state system (or three jars). For a brief description, the variable "a" stores a letter extracted at random from one of the components of the vector "jar" (from one of the three jars). This letter is returned to variable "a" by function "Draw". Before being called, function "Draw" may receive an integer variable "S" which specifies the component used

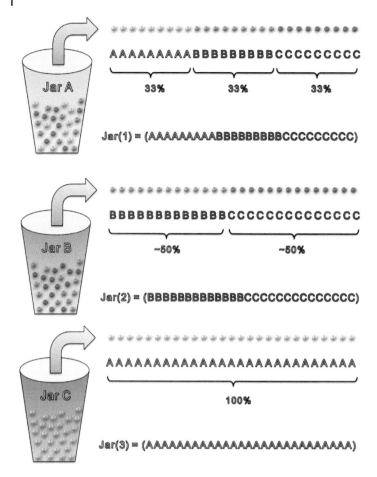

Figure 9.4 Translation of jars and balls to strings of letters. Each jar contains 27 balls of different colors. The balls from each jar are linearly arranged by color in a string (the order does not matter). The string of balls is replaced with a string of letters. Each string of letters is then stored in a vector designated as "Jar".

(the jar used). Each component of the vector "Jar" stores one string of 27 letters. In order to simulate a draw, function "Draw" generates a random integer between 1 and 27 that represents the letter found at *i*th position on a string of 27 letters. Note that vector "Jar" is declared globally, so it can be read both by the main function and other functions. Thus, assume the blue jar (state) is the first from which a draw is made. The blue jar is represented by component "Jar(2)" and a draw is simulated by using "Draw" function. Jar(2) contains two types of letters in equal proportions, namely "B" and "C". Therefore, the integer variable "*S*" of the function "Draw" takes "2" as the first value. If for instance "Draw" function returns letter "B" to variable "*a*", then the next draw is made again from the blue jar "B" (in our case, represented by component "Jar(2)"). If the function returns letter "C" to variable "*a*", then the next draw is made from the red jar "C" (represented by component "Jar(3)"). Since Jar(3) component stores only "A" letters, any letter that is chosen at random will point to the orange jar "A" (represented by component "Jar(1)"). Thus, this process continues until the 20th draw, and variable "*a*" content is recorded at every step in to a sequence of observations which represents the system behavior.

9.2.5.1 The Parameters for *n*-States

The implementation below allows the simulation of an unlimited number of states provided an extension of matrix "*P*" elements and vector "Jar" components depend on the situation. Matrix "*P*" stores the transition probabilities from one state to another. Vector "Jar" stores the string representations of the jars, whose proportion of letters reflects the transition probabilities stored in matrix "*P*". For an unlimited number of states, the transition matrix and vector "Jar" are declared as follows:

```
Dim P(0 To n, 0 To n-1) As Variant
Dim Jar(1 To n) As Variant
```

where *n* represents the number of states that are used for simulation. In this case, *n* equals three since there is a total of three states ($n = 3$). Note that all elements of matrix "*P*" are declared as "variant" (language-specific variable), which allows both the storage of numbers and letters. Thus, an immediate observation would be that rows are declared first and columns are declared second: *P*(row, column), which would be equivalent to the following meaning: *P*("jar *n*", "contains balls type $n - 1$") = "the proportion of balls". When it comes to implementation, certain optimizations must be made. Note that in the implementation case below, matrix "*P*" is not a square matrix (Figure 9.5). In the

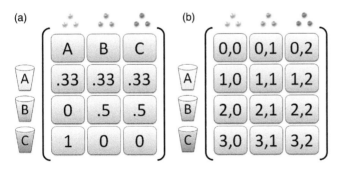

Figure 9.5 The mixed format of the matrix. The first row of the matrix (or the identity row of the states) is reserved for letters representative of states. The other rows store the transition probabilities. (a)Shows the format of matrix "*P*" from the algorithm implementation. (b) Shows the coordinates of the matrix "*P*" elements and how it relates to the implementation.

current example, the first row of the matrix "*P*" is reserved for the letters representative of the states used ($P(0, 0)$ = "A", $P(0, 1)$ = "B", $P(0, 2)$ = "C"). Thus, the first row of the matrix "*P*" is called the identity row of the states. The letters from the identity row are used in the first instance to generate the content for the components of vector "Jar". In the second instance, the letters stored in the first row of matrix "*P*" are used iteratively for the chain formation. This means that every letter that is returned by function "Draw" is compared to each of the letters stored in the first row of matrix "*P*". This comparison is made in order to select from which component (of the "Jar" vector) the next draw is made.

```
Dim P(0 To 3, 0 To 2) As Variant
Dim Jar(1 To 3) As Variant

Private Sub Form_Load()

P(0, 0) = "A"
P(0, 1) = "B"
P(0, 2) = "C"

P(1, 0) = 0.33
P(1, 1) = 0.33
P(1, 2) = 0.33

P(2, 0) = 0
P(2, 1) = 0.5
P(2, 2) = 0.5

P(3, 0) = 1
P(3, 1) = 0
P(3, 2) = 0

For j = 1 To 3
    Jar(j) = Fill_Jar(j)
    MsgBox Jar(j)
Next j

draws = 20
a = Draw(1)

For i = 1 To draws
    For j = 0 To 3
        If a = P(0, j) Then
            a = Draw(j + 1)
            q = q & P(0, j)
```

```
                  z = z & ", Jar " & P(0, j) & "[" & a & "]"
                  GoTo 1
             End If
      Next j
 1:

 Next i

 MsgBox q
 MsgBox z
 End Sub


 Function Fill_Jar(ByVal S As Variant) As Variant

 Ltot = 27

 For i = 0 To 2
      a = Int(Ltot * P(S, i))
            For j = 1 To a
                  b = b & P(0, i)
            Next j
 Next i
 Fill_Jar = b

 End Function

 Function Draw(ByVal S As Variant) As Variant
      Randomize
      randomly_choose = Int(Rnd * Len(Jar(S)))
      ball = Mid(Jar(S), randomly_choose + 1, 1)
      Draw = ball
 End Function
```

```
Output:
Q = CABCABCABCAABBBCABBB
Z = "Jar C[A], Jar A[B], Jar B[C], Jar C[A], Jar A[B], Jar B[C],
Jar C[A], Jar A[B], Jar B[C], Jar C[A], Jar A[A], Jar A[B], Jar
B[B], Jar B[B], Jar B[C], Jar C[A], Jar A[B], Jar B[B], Jar
B[B], Jar B[C]"
```

Supporting algorithm 14: A three-state Markov chain simulator. The probability values present inside a 3 × 3 transition matrix (*P*) are directly used for an automatic generation of the letter combination that make up the representation of the jars. Thus, the three letter sequences have a calculated proportion of "A", "B", and "C" letters. The chance of a letter chosen at random from one of the three sequences is directly dictated by the proportions of "A", "B", and "C" letters.

Nevertheless, the simulator shown above goes through two main stages: (1) it generates three strings of letters representative of three jars according to the matrix "*P*", (2) the strings of letters are used in the extraction under the chain model. At the end of the experiment, variable "*q*" contains 20 observations. Variable "*z*" contains the same observations as variable "*q*"; however, it stores all the outcomes in a more intuitive manner. The meaning of this type of output is: "From **Jar C [** ball **A** was randomly chosen**]**; since ball "A" was chosen, the extraction of the next ball is done from **Jar A [** where ball **B** was randomly chosen**]**, ..."

9.2.5.2   Automatic Generation of the Jar Content

There are situations in which the automatic generation of letter sequences for representing jars is highly needed for two reasons: (1) A large number of states, which will give rise to a large number of vector components. (2) The fair balance of chances when draws are made (if a constant number of letters is used for representing each jar). *The first reason* includes a large number of states. Suppose for a moment that instead of three states ($n = 3$) there are five hundred states ($n = 500$). This implies a state diagram composed of 500 states (or jars). To simulate 500 jars, each with its proportion of balls, a vector "Jar" with 500 components is needed (Jar(1 to 500)). Each component of this vector must contain a string of letters which is representative of the proportion of balls from a correspondent jar. An extreme case would be to consider that in a state diagram of 500 states, 499 of them communicate (send arrows to) with the remaining one; then the system would require a minimum of 500 different signs (since there are 24 letters in the alphabet, this would complicate the situation). *The second reason* involves the fair balance of chances when draws are made (see Section 9.2.4). In order to bring the odds closer to the ideal probability values of a transition matrix, the jars must be simulated by using long strings of letters. Therefore, the jar representation in the form of strings can be constructed by the computer with a specialized function. Such a primitive function ("Fill_Jar") was presented in Chapter 2 (see Section 2.3.4). There, the "Fill_Jar" function received two parameters: (1) The index of the vector component "*S*" and (2) a probability value "*p*" of one of the letters for the string of that particular component. Since the function was limited to only two states, the probability value of the complementary letter was determined by $1 - p$. In contrast, in order to represent a large number of states (from 2 to *n* states), the function presented here uses the transition probabilities directly from the transition matrix "*P*". Therefore, returning to our example of 27-letter representation, the implementation above uses a modified function ("Fill Jar") that is responsible for building the strings that are stored in each of the vector components. The function "Fill_Jar" receives in this case just one parameter, namely "*S*". This parameter is an integer variable which specifies the component to be filled with suitable proportions of letters "A", "B", or "C". Since there are only three states (three jars), the integer variable "*S*" takes 0, 1, and 2 as possible values. Initially, once the "Fill_Jar" function is called, the total number of letters (L_{tot}) is declared as a constant ($L_{tot} = 27$) for all three components involved (i.e., Jar(j), where j = 1, 2, 3). Next, the total number of letters is multiplied by transition probability values associated with a jar (from the corresponding row). This step is made in order to determine how many letters of a given type ("A", "B", or "C") are needed for the jar representation ($a = L_{tot} \times P(S, i)$). Depending on the results of these calculations, individual letters of each given type are attached one at a time to a string that is built and stored into variable "*b*" (see the above algorithm implementation). Then, the contents of the "b" variable are returned and stored in one of the components of the vector for future use. Once the "Fill_Jar" function was called for each component of the vector "Jar", the draw simulation begins.

9.3 Simulation of Different Chain Configurations

In one way or the other the behavior of the simulation must be tested. But what can be tested ? After simulating 20 draws, the algorithm provides an output of 20 observations (or 20 letters). One type of test would be the analysis of the frequency of each type of letter in the output. Thus, the frequency analysis will indicate the *actual average time* for each state. In Chapter 8, a number of Markov chain configurations have been discussed (Table 8.1). For each case, the *expected average time* was determined using a calculation method (see Section 7.4.2). Therefore, the test consists of a comparison between the *actual average time* from simulation and the *expected average time* from prediction.

9.3.1 Alternative Parameters and the Framework Expansion

However, in order to simulate each of these configurations, the parameters of the algorithm implementation must be adjusted according to the *number of states* and the *relations between states* (see Section 9.2.5.1). For an unlimited *number of states*, the transition matrix "*P*" and vector "*Jar*" are declared as follows:

```
Dim P(0 To n, 0 To n-1) As Variant
Dim Jar(1 To n) As Variant
```

where *n* represents the number of states that are used for simulation. The number of states *n* can be set to four since all the configurations discussed in Chapter 8 did not have more than four states ($n = 4$). Thus, the new form of the "Jar" vector and the matrix framework (*P*) may be written as:

```
Dim P(0 To 4, 0 To 3) As Variant
Dim Jar(1 To 4) As Variant
```

Table 9.2 Alternative parameters of the transition matrix for cases A, L, F, and G. The configuration of diagrams is shown in the first row of the table. The transition matrix of each diagram is declared vertically in a linear format.

	Case A	Case L	Case F	Case G
A	P(1, 0) = 0	P(1, 0) = 1	P(1, 0) = 0	P(1, 0) = 0
	P(1, 1) = 1	P(1, 1) = 0	P(1, 1) = 0.5	P(1, 1) = 1
	P(1, 2) = 0	P(1, 2) = 0	P(1, 2) = 0.5	P(1, 2) = 0
	P(1, 3) = 0	P(1, 3) = 0	P(1, 3) = 0	P(1, 3) = 0
B	P(2, 0) = 0.5	P(2, 0) = 0.5	P(2, 0) = 1	P(2, 0) = 0.33
	P(2, 1) = 0.5	P(2, 1) = 0	P(2, 1) = 0	P(2, 1) = 0
	P(2, 2) = 0	P(2, 2) = 0.5	P(2, 2) = 0	P(2, 2) = 0.33
	P(2, 3) = 0	P(2, 3) = 0	P(2, 3) = 0	P(2, 3) = 0.33
C	P(3, 0) = 0	P(3, 0) = 0	P(3, 0) = 0	P(3, 0) = 0
	P(3, 1) = 0	P(3, 1) = 0.5	P(3, 1) = 0	P(3, 1) = 1
	P(3, 2) = 0	P(3, 2) = 0	P(3, 2) = 0	P(3, 2) = 0
	P(3, 3) = 0	P(3, 3) = 0.5	P(3, 3) = 1	P(3, 3) = 0
D	P(4, 0) = 0	P(4, 0) = 0	P(4, 0) = 0	P(4, 0) = 0
	P(4, 1) = 0	P(4, 1) = 0	P(4, 1) = 0.5	P(4, 1) = 0
	P(4, 2) = 0	P(4, 2) = 1	P(4, 2) = 0.5	P(4, 2) = 1
	P(4, 3) = 0	P(4, 3) = 0	P(4, 3) = 0	P(4, 3) = 0

Note that all the elements of matrix "*P*" are declared as "Variant" (language-specific variable), which allows both the storage of numbers and letters (see Section 9.2.5.1). The *relations between states* are contained within the transition matrix "*P*" as transition probability values. Thus, an immediate observation would be that rows are declared first and columns are declared second: $P(row, column)$, which would be equivalent to the following meaning: **P("jar *n*", "contains balls type *n*-1")="probability value"**. Since $n = 4$, the values of 16 elements of the transition matrix must be declared beforehand. Thus, the second modification is represented by the addition of probability values. Statements are divided into four groups, each representing a row of transition matrix "*P*" (or each representing a jar). For instance, the implementation below has all the parameters set for the study of case G (Figure 8.7 and Table 9.2):

```
Dim P(0 To 4, 0 To 3) As Variant
Dim Jar(1 To 4) As Variant

Private Sub Form_Load()

P(0, 0) = "A"
P(0, 1) = "B"
P(0, 2) = "C"
P(0, 3) = "D"

P(1, 0) = 0
P(1, 1) = 1
P(1, 2) = 0
P(1, 3) = 0

P(2, 0) = 0.33
P(2, 1) = 0
P(2, 2) = 0.33
P(2, 3) = 0.33

P(3, 0) = 0
P(3, 1) = 1
P(3, 2) = 0
P(3, 3) = 0

P(4, 0) = 0
P(4, 1) = 0
P(4, 2) = 1
P(4, 3) = 0

For j = 1 To 4
    Jar(j) = Fill_Jar(j)
Next j

draws = 100
a = Draw(1)

For i = 1 To draws
    For j = 0 To 3
        If a = P(0, j) Then
            a = Draw(j + 1)
            z = z & P(0, j)
```

```
            GoTo 1
        End If
    Next j
1:
Next i

MsgBox z

End Sub

Function Fill_Jar(ByVal S As Variant) As Variant

Ltot = 100

For i = 0 To 3
    a = Int(Ltot * P(S, i))
        For j = 1 To a
            b = b & P(0, i)
        Next j
Next i

Fill_Jar = b
End Function

Function Draw(ByVal S As Variant) As Variant
    Randomize
    randomly_choose = Int(Rnd * Len(Jar(S)))
    ball = Mid(Jar(S), randomly_choose + 1, 1)
    Draw = ball
End Function
```

```
Output:
Z=BABABDCBABDCBCBDCBDCBABDCBCBCBCBDCBCBDCBDCBDCBABABABDCBDCBABAB
DCBABCBABCBABDCBDCBABABABABCBABCBDCBDC
```

Supporting algorithm 15: A Markov chain framework for simulation. The probability values present inside a 4 × 4 transition matrix (*P*) are directly used for an automatic generation of the letter combination that make up the representation of four jars. Thus, the four letter sequences have a calculated proportion of "A", "B", "C", and "D" letters. The chance of a letter chosen at random from one of the four sequences is directly dictated by the proportions of "A", "B", "C", and "D" letters.

Thus, in the above algorithm, the transition matrix "*P*" of case G is firmly declared in a linear format. Compared to the previous implementation, the output is shown only as a series of observations (of letters). However, the aim of the above algorithm is to allow the simulation of a wide number of diagrams.

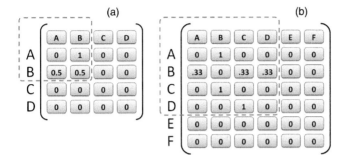

Figure 9.6 Framework expansion. It shows that matrix frameworks of *n* states can be used for a wide range of diagram configurations, ranging between 2 and *n* states. (a) The orange frame shows the transition matrix of a diagram (case A example) of two states used on a matrix framework of four states. (b) In this panel, the orange frame shows the transition matrix of another diagram (case G example) of four states used on a matrix framework of six states.

What can be changed in order to mold the algorithm in a new situation, a new diagram? For example, the configuration of diagrams for cases A, L, F, and G is shown in the first line of Table 9.2, and the transition matrix of each diagram is declared vertically in a linear format. Instead of using case G diagram as before, suppose that case F diagram is used for simulation (Table 9.2). Case F has the same number of states as case G and therefore, a transition matrix with the same number of elements. Thus, simply by changing the parameters of the algorithm to the new transition probabilities of case F, the behavior of case F diagram can be simulated accordingly. Diagrams with fewer than four states can also be simulated without additional changes of the above implementation (Figure 9.6a). Notice that although diagram of case A has only two states, the transition matrix corresponding to this diagram is molded on the same matrix framework of four rows and four columns (Table 9.2 and Figure 9.6a). Here, the first two rows and two columns in the framework will represent the two-state system and all other elements of the matrix framework are filled with zero values. Thus, the simulation algorithm will neglect the other statements that have zero transition probabilities (Figure 9.6). Therefore, by simply modifying the transition probabilities from the above implementation accordingly, any diagram configuration ranging from two to four states can be simulated. Moreover, in this framework, any of the configurations discussed so far can be tested. However, the implementation is limited to a maximum of four states ($n = 4$). For an expansion of this framework, larger matrices can be used. For instance, matrices of six states ($n = 6$) instead of four states ($n = 4$) can be built (Figure 9.6b). Thus, one can test diagram configurations of two states up to six states using the same implementation above. The first step in this expansion/modification of the simulator is represented by a redeclaration of the "Jar" vector and the matrix framework (composed of the identity row and a square matrix). Therefore, the new form of the "Jar" vector and the matrix framework may be written as:

```
Dim P(0 To 6, 0 To 5) As Variant
Dim Jar(1 To 6) As Variant
```

The second step in this expansion is the declaration of each element of the matrix framework in a linear manner, similar to what is shown in Table 9.2. As described previously, the identity row takes the form of **P(0, n)="letter"** and represents the first group of statements. The remaining statements consist of six other groups. Each of the remaining groups represents a line of the transition matrix "*P*", and can be interpreted as: **P("State *n*", "goes to state *n*-1")="with a probability value of".** Nevertheless, if a diagram of four states (case G for instance) must be tested in a framework of six states, then some of the elements of the matrix framework will be unused and equal to zero (Figure 9.6b).

9.3.2   The Output Verification

Up to this point, variants of the simulation have been discussed on numerous diagram configurations, in which the output is represented by a string of letters. The string of letters is representative of the diagram behavior. Thus, if the string of letters from the output is analyzed, the transition probabilities between letters should reflect the transition probabilities of the initial transition matrix. In other words, one way to test the correct functionality of the simulator is the rebuilding of the initial transition matrix by analyzing the output letters. Then a comparison between the two transition matrices should show a similarity between each counterpart element (i.e., $T1[m_{11}]$ should have close values to $T2[m_{11}]$). The output of the previous implementation mimics the behavior of case G. Therefore, a test is made on the previous output (Supporting algorithm 15) which provided 100 observations (letters):

```
"BABABDCBABDCBCBDCBDCBABDCBCBCBCBDCBCBDCBDCBDCBABABABDCBDCBABAB
DCBABCBABCBABDCBDCBABABABABCBABCBDCBDC"
```

Methods described previously (see Chapters 2 and 3) use a standard formula for calculating the transition probabilities between letters. This standardized formula is used here for the same purpose. First, the pairs of letters (transitions from state to state) are counted in order to extract the transition probabilities from the output of Supporting algorithm 15. Given that there are four states involved (A, B, C, and D), a total of 16 transitions ($4 \times 4 = 16$) should be calculated. As shown in Chapter 2, the transition probabilities can be calculated as follows:

$$T_{a \to b} = \frac{\text{Count } (D_{a \to b})}{\text{Count } (N_a)}$$

where $T_{a \to b}$ is the transition probability from letters that represent a, to letters that represent b. $D_{a \to b}$ represents the number of transitions from a to b and N_a represents the number of times a appears in the sequence. For instance, if the transition probability from "B" to "D" (which represents just one of the 16 transition probabilities) must be found in the above sequence, then a is represented by the letter "B" and b is represented by the letter "D". In this case, $D_{a \to b}$ counts how many times "BD" appears in the sequence, and N_a counts how many times "B" appears in the sequence. If the number of transitions between "B" and "C" is highlighted using a bold font, the counting of the letters can be made slightly easier:

"B̶ABA**BD**CBA**BD**CBC**BD**C**BD**CBA**BD**CBCBCBC**BD**CBC**BD**C**BD**C**BD**CBABABA**BD**C**BD**CBABAB DCBABCBABCBA**BD**C**BD**CBABABABABCBABC**BD**C**BD**C̶"

As noted above, whether it is a "B" or a "D" state, the counting of states is made from the second letter and ends at the penultimate letter in the sequence. Notice that since they are not counted, the first and the last letter in the sequence are represented with a strikethrough line (see Chapter 2). The underline shows the individual letters (N_a) and the transitions between letters ($D_{a \to b}$) are shown in bold. The only transition which is not counted is between the first and the second letter (state). Thus, the count shows that the pair of letters "BD" occurs 16 times ($D_{B \to D} = 16$) and individual letter "B" appears 41 times ($N_B = 41$). The probability of transition between the two letters can be found by replacing the values in the above formula:

$$T_{B \to D} = \frac{\text{Count } (D_{B \to D})}{\text{Count } (N_B)} = \frac{16}{41} = 0.39$$

Once the above formula has been used for calculating all transition probabilities (all 16 transitions), a transition matrix can be built. Using the same rationale, such calculations can be made automatically using a specialized function. An implementation that builds a transition matrix based on the output of the previous implementation is shown below.

```
Dim P(1 To 4, 1 To 4) As String

Private Sub Form_Load()

Call ExtractProb(
"BABABDCBABDCBCBDCBDCBABDCBCBCBCBDCBCBDCBDCBDCBABA" & _
"BABDCBDCBABABDCBABCBABCBABDCBDCBABABABABCBABCBDCBDC")
```

```
For i = 1 To 4
    For j = 1 To 4
        z = z & Chr(9) & Round(P(i, j), 2)
    Next j
    z = z & vbCrLf
Next i

MsgBox z
End Sub

Function ExtractProb(ByVal s As String)

Ea = "A"
Eb = "B"
Ec = "C"
Ed = "D"

For i = 1 To 4
    For j = 1 To 4
      P(i, j) = 0
    Next j
Next i

Ta = 0
Tb = 0
Tc = 0
Td = 0

For i = 2 To Len(s) - 1

        DI1 = Mid(s, i, 1)
        DI2 = Mid(s, i + 1, 1)

        If DI1 = Ea Then r = 1
        If DI1 = Eb Then r = 2
        If DI1 = Ec Then r = 3
        If DI1 = Ed Then r = 4

        If DI2 = Ea Then c = 1
        If DI2 = Eb Then c = 2
        If DI2 = Ec Then c = 3
        If DI2 = Ed Then c = 4
```

```
       P(r, c) = Val(P(r, c)) + 1

       If DI1 = Ea Then Ta = Ta + 1
       If DI1 = Eb Then Tb = Tb + 1
       If DI1 = Ec Then Tc = Tc + 1
       If DI1 = Ed Then Td = Td + 1

Next i

For i = 1 To 4
    For j = 1 To 4
        If i = 1 Then P(i, j) = Val(P(i, j)) / Ta
        If i = 2 Then P(i, j) = Val(P(i, j)) / Tb
        If i = 3 Then P(i, j) = Val(P(i, j)) / Tc
        If i = 4 Then P(i, j) = Val(P(i, j)) / Td
    Next j
Next i

End Function
```

```
Output:
         0      1    0      0
         0.39   0    0.22   0.39
         0      1    0      0
         0      0    1      0
```

Supporting algorithm 16: Transition probability tester. Previously, a sequence of observations has been provided by a simulator. To test the accuracy of the simulator, the sequence of observations is used for creating a transition matrix, which is then compared with the original.

The algorithm is based on a main function called "ExtractProb". In the first cycle, this function counts the pairs of letters and stores them in a matrix form. The counts of individual letters are stored in separate variables (Ta, Tb, Tc, and Td). In the second cycle, the formula for $T_{a \to b}$ is applied by dividing the number of pairs of letters from the matrix to the counts of individual letters. The output of the function is shown in the above implementation under the form of a transition matrix ($T2$). Thus, when the transition matrix $T1$ (originally used as a parameter for simulation) and transition matrix $T2$ (built from the output of the simulation) are placed side by side, a visual comparison can be performed (Figure 9.7). Figures 9.7a and 9.7b shows that the two square matrices ($T1$ and $T2$) contain close transition probability values. Thus, the test indicates that the algorithm of the simulator performs as expected. If the simulator is turned on again and another output of 100 letters is analyzed, the transition probability

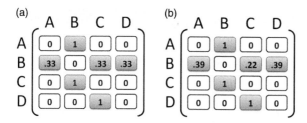

Figure 9.7 The verification of the output. (a) Initial transition matrix (*T1*) of the simulator. (b) Transition probabilities between letters of the output presented in the form of a transition matrix (*T2*).

values of matrix *T2* will vary slightly but not too far from the initial transition probability values of matrix *T1*. However, the two matrices will be more and more similar as the string of observations becomes longer (i.e., 1000 letters instead of 100 letters). Thus, the transition probability values of matrix *T2* will converge increasingly more toward the transition probability values of matrix *T2* as the sequence of observations tends to infinity.

9.3.3 Determination of Average Time by Experiment

So far several stages have been covered. First, some discussions have been made about the design of the simulator and then some discussions related to the verification of functionality. One of the utilities of the simulator is the determination of the average time by experiment. Therefore, the test consists of a comparison between the *actual average time* from simulation and the *expected average time* from prediction.

9.3.3.1 The Analysis of Frequency

To perform the final experiment, some additions to the previous implementation must be made. Instead of just showing the output, this time the output must be silently analyzed letter by letter. First, for an accurate experiment, the output of the simulator (number of draws) must be increased considerably. Thus, the implementation below uses an output of 10,000 letters instead of 100 letters. Second, the frequency of each letter is determined using the following formula:

$$F_N = \frac{100}{L_{out}} \times \text{Count(N)}$$

where N represents one of the letters associated with a state (i.e., A, B, C, or D), and L_{out} is the total length of the output (in this case, $L_{out} = 10,000$). The frequency (F_N) is representative of the average time spent in a state and it is expressed in percentages. For instance, suppose that in a string of 10,000 letters, there are 1500 letters type "A". Thus, the frequency of letter "A" (F_A) will be:

$$F_A = \frac{100}{L_{out}} \times \text{Count(A)} = \frac{100}{10,000} \times 1500 = 15\%$$

From this single example, it can be observed that this hypothetical system spends about 15% of the time in state "A". Therefore, the frequency analysis of all four letter types (F_A, F_B, F_C, and F_D) will indicate the *actual average time* for each state. The implementation below simulates the state diagram of case G. First, it generates an output of 10,000 observations (of letters). Next, the frequency of each letter is calculated based on these observations and is presented as the final result.

```
Dim P(0 To 4, 0 To 3) As Variant
Dim Jar(1 To 4) As Variant
Dim f(0 To 3) As Variant


Private Sub Form_Load()


P(0, 0) = "A"
P(0, 1) = "B"
P(0, 2) = "C"
P(0, 3) = "D"


P(1, 0) = 0
P(1, 1) = 1
P(1, 2) = 0
P(1, 3) = 0


P(2, 0) = 0.33
P(2, 1) = 0
P(2, 2) = 0.33
P(2, 3) = 0.33


P(3, 0) = 0
P(3, 1) = 1
P(3, 2) = 0
P(3, 3) = 0


P(4, 0) = 0
P(4, 1) = 0
P(4, 2) = 1
P(4, 3) = 0
```

```
For j = 1 To 4
    Jar(j) = Fill_Jar(j)
Next j

draws = 10000
a = Draw(2)

For i = 1 To draws
    For j = 0 To 3
        If a = P(0, j) Then
            a = Draw(j + 1)
            z = z & a
            GoTo 1
        End If
    Next j
1:
Next i

For i = 1 To Len(z)
    g = Mid(z, i, 1)
    If g = "A" Then f(0) = f(0) + 1
    If g = "B" Then f(1) = f(1) + 1
    If g = "C" Then f(2) = f(2) + 1
    If g = "D" Then f(3) = f(3) + 1
Next i

For i = 0 To 3
pro = pro & P(0, i) & "=" & Int((100 / Len(z)) * f(i)) & "%"
& Chr(9)
Next i

MsgBox pro
End Sub

Function Fill_Jar(ByVal S As Variant) As Variant
Ltot = 100
For i = 0 To 3
    a = Int(Ltot * P(S, i))
        For j = 1 To a
            b = b & P(0, i)
        Next j
Next i
```

```
Fill_Jar = b
End Function


Function Draw(ByVal S As Variant) As Variant
    Randomize
    randomly_choose = Int(Rnd * Len(Jar(S)))
    ball = Mid(Jar(S), randomly_choose + 1, 1)
    Draw = ball
End Function
```

```
Output:
State A=15%    State B=42%    State C=26%    State D=15%
```

Supporting algorithm 17: Average time tester. The tester is composed of a simulator that generates 10,000 observations. These observations are then analyzed and the frequencies of letters "A", "B", "C", and "D" are determined. These frequencies represent the average time spent in each state.

The observations produced by the simulator are stored in variable "z" (Supporting algorithm 17). The content of variable "z" is then analyzed and the letters are counted. The counting of the four letters is made using a vector "f" with four components. The letter frequency is shown directly in the output by using the expression discussed above, namely: $(100 / L_{out}) \times f(i)$, where i represents the index of one of the four components of vector "f", and L_{out} is the total length of the observations produced by the simulator. Therefore, the frequency of letters shows that state diagram of case G spends about 15% of the time in state "A", about 42% of the time in state "B", about 26% of the time in state "C", and about 15% of the time in state "D". Of course, since a random process is involved, these percentages are not fixed and may vary on each run of the algorithm. However, the variation is relatively small when the number of observations increases.

9.3.3.2 Frequency vs. Prediction

In this chapter, the main exemplification has been made on the state diagram of case G. Previously, the *expected average time* of case G was determined by using a calculation method (see Sections 7.3 and 7.4.2). However, here the calculation of the *expected average time* is made again in order to be compared with data originated from the simulator. First, the corresponding transition matrix (P) of case G state diagram is written as:

$$
P = \begin{pmatrix}
0.00 & 1.00 & 0.00 & 0.00 \\
0.33 & 0.00 & 0.33 & 0.33 \\
0.00 & 1.00 & 0.00 & 0.00 \\
0.00 & 0.00 & 1.00 & 0.00
\end{pmatrix}
$$

In order to determine the average time ($t(s)$) spent on each state, a transition matrix (m) is obtained from the multiplication of matrix (P) with itself:

$$m_{ij} = \sum_{k=1}^{q} P_{i,k} \times P_{k,j} = (P_{1,k} \times P_{k,1}) + (P_{2,k} \times P_{k,2}) + \cdots + (P_{i,q} \times P_{q,j})$$

Thus, each m_{ij} entry is given by multiplying the entries $P_{i,k}$ (across row i of P) by the entries of the same square matrix $P_{k,j}$ (across column j of P), for $k = 1$, 2, ..., q, and summing the results over k. Thus, the following transition matrix (m) can be written as:

$$m = P \times P = \begin{pmatrix} 0.33 & 0.00 & 0.33 & 0.33 \\ 0.00 & 0.66 & 0.33 & 0.00 \\ 0.33 & 0.00 & 0.33 & 0.33 \\ 0.00 & 1.00 & 0.00 & 0.00 \end{pmatrix}$$

The average time spent in a state (s) can be expressed in percentages as:

$$t(s) = \left(\sum_{i=1}^{q} m_{is} \right) \times \frac{100}{q}$$

where q represents the number of states, and s represents the columns of matrix m. Case G has four states (state "A", "B", "C", and "D"), therefore $q = 4$. Thus, the average time spent in state "A" ($s = 1$), state "B" ($s = 2$), state "C" ($s = 3$), and state "D" ($s = 4$), can be found using the above formula, namely:

$$t(1) = \left(\sum_{i=1}^{4} m_{i1} \right) \times \frac{100}{4} = (0.33 + 0 + 0.33 + 0) \times \frac{100}{4} = 17\%$$

$$t(2) = \left(\sum_{i=1}^{4} m_{i2} \right) \times \frac{100}{4} = (0 + 0.66 + 0 + 1) \times \frac{100}{4} = 41\%$$

$$t(3) = \left(\sum_{i=1}^{4} m_{i3} \right) \times \frac{100}{4} = (0.33 + 0.33 + 0.33 + 0) \times \frac{100}{4} = 25\%$$

$$t(4) = \left(\sum_{i=1}^{4} m_{i4} \right) \times \frac{100}{4} = (0.33 + 0 + 0.33 + 0) \times \frac{100}{4} = 17\%$$

As indicated by the above calculations, in a series of simulations, *the expectations* from the state diagram of case G are: about 17% of the time the system should be in state "A", approximately 41% of the time the system is expected to be in state "B", 25% of the time the system should be in state "C", and about 17% of the time the system should be in state "D" (Table 9.3). Data from the simulation indicate the presence of similar percentages. The frequency of letters shows that state diagram of case G actually spends: about 15% of the time in state "A",

Table 9.3 Frequency vs. prediction. The table concerns the state diagram of case G and it shows in a comparative manner the expected average time from prediction and the actual average time from simulation.

Case G	A	B	C	D
Predicted	17%	41%	25%	17%
Experiment	15%	42%	26%	15%
Average	16%	41.5%	25.5%	16%
SD	±1.4	±0.7	±0.7	±1.4

about 42% of the time in state "B", about 26% of the time in state "C", and about 15% of the time in state "D" (Table 9.3). Thus, the time spent in the state "A" and state "D" is approximately 2% below expectations. On the other hand, the time spent in state "B" and state "C" is 1% above expectations (Table 9.3). Of course, these percentages vary each time another experiment is made. Nevertheless, the data from this particular simulation indicate that the results obtained from the prediction are reliable and supported by the experiment.

9.3.3.3 Average Time by Experiment on Various State Diagrams

A step-by-step analysis has been made for the state diagram of case G. However, many other configurations of state diagrams can be tested in the same manner. Further examples can be considered in other five cases, namely: case D, case E, case F, case G, and case H. The predictions for the mentioned cases have been already made in Chapter 8. Thus, those predicted values can be used comparatively with the data from the simulator (Table 9.4). Table 9.4 suggests that results obtained from the simulator are consistent with the predicted values. For each of the five experiments, a total 10,000 observations have been used. With exactly the same parameters on each simulation, the *actual average time* may vary within certain limits. Nevertheless, up to this point, a simulator for *n*-state diagrams was designed, built and tested.

9.3.4 A Window into the Behavior of Random Processes

All determinations made so far consisted in the analysis of the final output generated by the simulator. Thus, the frequency of each letter was calculated only after the algorithm recorded all the observations inside one final output. Therefore, the determination made is only a slice of the system behavior. Of course, this slice provides accurate information on the convergence of the letter frequency as the output increases. However, such a slice does not provide

Table 9.4 Observations vs. expectations. The left side of the table shows the state diagrams of: case D, case E, case F, case G, and case H. The right side of the table shows a comparison between actual average time (in dark red) from simulation and the expected average time (in orange) from prediction.

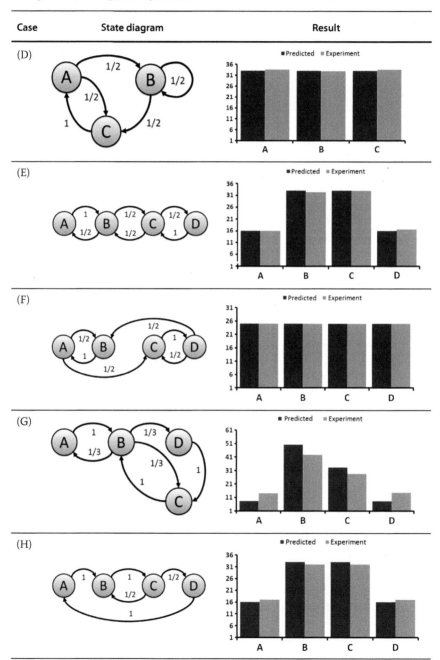

information about the system behavior over time. What happens if the letter frequency is analyzed only after 500 observations? How about after 1000 observations or 10,000 observations? Whether a frequency analysis is done at 500, 1000, or 10,000 observations, the information obtained consists of three separate slices. If the frequencies of the three slices are analyzed in a single context, then a convergence of the frequency of each individual letter can be observed. In this manner, a general overview of the system behavior can be made. However, what if the letter frequency is analyzed after each observation is made (every time the chain of observations increases)? In this case, there is a slice for each step that the system makes over time. In other words, the general situation of the machine can be presented at each discrete step. For instance, if a total of 1000 observations are made, then there will be a total of 1000 slices whose letter frequencies can be plotted on a graph (Figure 9.8). Thus, Figure 9.8 shows the long-term behavior of the state diagram of case G. Here, two slices are presented for illustration. The first slice shows the frequency of letters after 250 observations and the second slice shows the letter frequency after 450 observations. The information from the first slice indicates that state "A" was visited less often than state "D" after precisely 250 steps. However, the information from the second slice indicates that state "A" was visited almost as often as state "D" after 450 steps. This trend may decrease or increase with the evolution of the system over time (Figure 9.8). Thus, the distribution suggests that

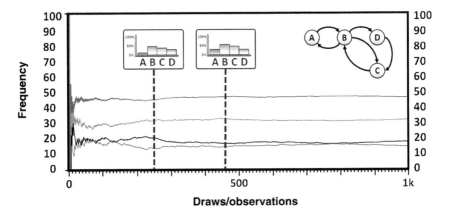

Draws/observations

Figure 9.8 A window into random processes. Shows the behavior of state diagram from case G on 1000 discrete steps. x-axis indicates the letter frequency for each discrete step as follows: frequency of letter "A" is shown by the red line, frequency of letter "B" is shown by the blue line, frequency of letter "C" is shown by the orange line, and the frequency of letter "D" is shown by the black line. The y-axis represents the number of discrete steps made by the system. The two charts from within the distribution show the frequency of each letter for 250 steps and for 450 steps.

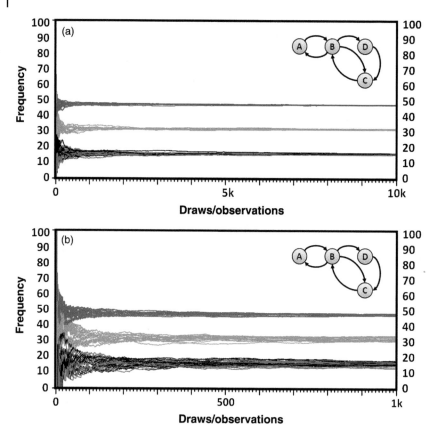

Figure 9.9 A window into randomness. The figure shows the overlapping distribution of a series of experiments made on the state diagram of case G. (a) It shows the distribution of a total of 20 separate experiments on 10,000 discrete steps, (b) shows an enlarged distribution of a distinct set of 20 experiments made on 1000 discrete steps. In both graphs, the frequency of letter "A" is shown by a red line, the frequency of letter "B" is shown by the blue line, the frequency of letter "C" is shown by the orange line, and the frequency of letter "D" is shown by the black line.

in short time periods, the behavior of the system is unpredictable (Figure 9.8). Nevertheless, in longer time periods (more discrete steps), the behavior of the system converges to an expected average (Figure 9.9a). In other words, the frequency of each letter revolves around an average and the variation decreases. Now, every time the simulator is "switched on" with exactly the same parameters, the behavior of the system varies in detail. Thus, the distribution on 1000 steps is different each time. Perhaps the best method to highlight the convergence toward the expected average consists of a superimposed distribution of repeated experiments (Figures 9.9a and 9.9b). Thus, Figure 9.9 shows the many pathways of a randomized process, namely a superimposed distribution of 20 separate experiments. Although the parameters (transition probabilities) of the simulator dictate the final outcome, a random variation of the letter frequency can be observed on the y-axis. However, the effect of the random variation on the outcome is less and less visible as the system evolves over time (Figure 9.9).

A

Supporting Algorithms in PHP

Supporting algorithm (PHP alternative) 1.

```php
<?php

$Jar = array();
$draws = 17;

$Jar[0] = "WWBBBBBBBB";
$Jar[1] = "WWWWWBBBBB";

$a = Draw(0);
$z = $z . " Jar W[" . $a . "],";

for ($i=1; $i<=$draws; $i++){
    If ($a == "W") {
        $a = Draw(0);
        $z = $z . " Jar W[". $a . "],";
    } else {
        $a = Draw(1);
        $z = $z . " Jar B[" . $a . "],";
    }
}

echo $z;

function Draw($S) {
    srand(mktime());
    global $Jar;
    $randomly_choose = rand(1,strlen($Jar[$S])-1);
    $ball = substr($Jar[$S], $randomly_choose, 1);
```

Markov Chains: From Theory to Implementation and Experimentation, First Edition. Paul A. Gagniuc.
© 2017 John Wiley & Sons, Inc. Published 2017 by John Wiley & Sons, Inc.
Companion website: www.wiley.com/go/gagniuc/markovchains

```
        return $ball;
}
?>
```

```
PHP Output:
Jar W[B], Jar B[W], Jar W[B], Jar B[B], Jar B[B], Jar B[W],
Jar W[B], Jar B[B], Jar B[W], Jar W[W], Jar W[B], Jar B[B],
Jar B[B], Jar B[B], Jar B[B], Jar B[W], Jar W[W], Jar W[B],
```

Supporting algorithm (PHP alternative) 2.

```php
<?php
$Jar = array();
$draws = 17;

Fill_Jar(0, 0.2);
Fill_Jar(1, 0.6);

$a = Draw(0);
$z = $z . " Jar W[" . $a . "],";

for ($i=1; $i<=$draws; $i++){

    If ($a == "W") {
        $a = Draw(0);
        $z = $z . " Jar W[" . $a . "],";
    } else {
        $a = Draw(1);
        $z = $z . " Jar B[" . $a . "],";
    }
}
echo $z;

function Draw($S) {
    srand(mktime());
    global $Jar;
    $randomly_choose = rand(1,strlen($Jar[$S])-1);
    $ball = substr($Jar[$S], $randomly_choose, 1);
    return $ball;
}

function Fill_Jar($S, $p){
```

```
global $Jar;

$Balls_W = round(100 * $p);
$Balls_B = 100 - $Balls_W;


for ($i=1; $i<=$Balls_W; $i++){
    $Jar[$S] = $Jar[$S] . "W";
}

for ($i=1; $i<=$Balls_B; $i++){
    $Jar[$S] = $Jar[$S] . "B";
}
}
?>
```

```
PHP Output:
Jar W[B], Jar B[W], Jar W[B], Jar B[B], Jar B[W], Jar W[B],
Jar B[W], Jar W[B], Jar B[W], Jar W[B], Jar B[W], Jar W[B],
Jar B[W], Jar W[W], Jar W[B], Jar B[W], Jar W[B], Jar B[B],
```

Supporting algorithm (PHP alternative) 3.

```
<?php

$M = array();

ExtractProb("SRRSRSRRSRSRRSS");

Function ExtractProb($s) {
global $M;

$Eb = "S";
$Es = "R";

for($i=1; $i<=2; $i++){
    for($j=1; $j<=2; $j++){
        $M[$i][$j]=0;
    }
}

$TB = 0;
$TS = 0;
```

```
for($i=1; $i<strlen($s)-1; $i++){

        $DI1 = substr($s, $i, 1);
        $DI2 = substr($s, $i+1, 1);

        If ($DI1 == $Eb) {$r = 1;}
        If ($DI1 == $Es) {$r = 2;}
        If ($DI2 == $Eb) {$c = 1;}
        If ($DI2 == $Es) {$c = 2;}

        $M[$r][$c] = $M[$r][$c] + 1;

        If ($DI1 == $Eb) {$TB = $TB + 1;}
        If ($DI1 == $Es) {$TS = $TS + 1;}
}

echo DrowMatrix(2, 2, $M, "(C)", "Count:");

for($i=1; $i<=2; $i++){
    for($j=1; $j<=2; $j++){
        If ($i == 1) {$M[$i][$j] = intval($M[$i][$j]) / $TB;}
        If ($i == 2) {$M[$i][$j] = intval($M[$i][$j]) / $TS;}
    }
}

echo DrowMatrix(2, 2, $M, "(P)", "Transition matrix M:");
}

Function DrowMatrix($ib, $jb, $M, $model, $msg) {
$ct = "";
$Eb = "S";
$Es = "R";

$ct .= "<table border='1'><tr>";
$ct .= "<td>" . $model . "</td><td>" . $Eb . "</td><td>" .
$Es . "</td></tr>";

for($i=1; $i<=$ib; $i++){
        $ct .= "<tr>";
        for($j=1; $j<=$jb; $j++){
                $v = Round($M[$i][$j], 1);
                If ($j == 1 and $i == 1) {$ct .= "<td>" .
```

```
                    $Eb . "</td>";}
                    If ($j == 1 and $i == 2) {$ct .= "<td>" .
                    $Es . "</td>";}
                    $ct .= "<td>" . $v . "</td>";
            }
$ct .= "</tr>";
}

$ct .= "</table>";
$rez = $msg . " M[" . $jb . "," . $ib . "]" . "</br>" . $ct
. "</br>";

return $rez;
}
?>
```

```
PHP Output:
Count: M[2,2]
```

(C)	S	R
S	1	4
R	5	3

Transition matrix M: M[2,2]

(P)	S	R
S	0.2	0.8
R	0.6	0.4

Supporting algorithm (PHP alternative) 4.

```php
<?php
$P = array();
$v = array();

$chain = 5;

$P[1][1] = 0.2;
$P[1][2] = 0.625;
$P[2][1] = 0.8;
$P[2][2] = 0.375;

$v[0] = 1;
$v[1] = 0;
```

```
for($i=1; $i<=$chain;$i++){
    $x = ($v[0] * $P[1][1]) + ($v[1] * $P[1][2]);
    $y = ($v[0] * $P[2][1]) + ($v[1] * $P[2][2]);

    $v[0] = $x;
    $v[0] = $Y;

    echo $v[0] . " | " . $v[1] . "</br>";
}
?>
```

```
PHP Output:
0.2 | 0.8
0.54 | 0.46
0.3955 | 0.6045
0.4569125 | 0.5430875
0.4308121875 | 0.5691878125
```

Supporting algorithm (PHP alternative) 5.

```
<?php
$P = array();
$v = array();

$chain = 50;

$P[1][1] = 0.2;
$P[1][2] = 0.625;
$P[2][1] = 0.8;
$P[2][2] = 0.375;

$v[0] = 1;
$v[1] = 0;

for($i=1; $i<=$chain; $i++){
    $x = ($v[0] * $P[1][1]) + ($v[1] * $P[1][2]);
    $y = ($v[0] * $P[2][1]) + ($v[1] * $P[2][2]);

    if($v[0] == $x AND $v[1] == $y){
        echo "Steady state vector at day [" . $i . "]!";
        $i = $chain;
    } else {
```

```php
        echo "Day[" . $i . "], v=[" . $x . " | " . $y .
        "] </br>";
    }

    $v[0] = $x;
    $v[1] = $y;
}
?>
```

Supporting algorithm (PHP alternative) 6.

```php
<?php
$P = array();
$v = array();

$P[1][1] = 0.2;
$P[1][2] = 0.625;
$P[2][1] = 0.8;
$P[2][2] = 0.375;

$a = $P[1][1];
$b = $P[2][2];

$x = (1 - $b) / (2 - ($a + $b));
$y = (1 - $a) / (2 - ($a + $b));

$v[0] = $x;
$v[1] = $y;

echo "Steady state vector v = [" . $v[0] . " | " . $v[1] .
"]";
?>
```

```
PHP Output:
Steady state vector v = [0.43859649122807 | 0.56140350877193]
```

Supporting algorithm (PHP alternative) 7.

```php
<?php
$M = array();
$v = array();
```

```
ExtractProb("SRRSRSRRSRSRRSS");

$chain = 5;

$v[0] = 1;
$v[1] = 0;

for($i=1; $i<=$chain; $i++){
    $x = ($v[0] * $M[1][1]) + ($v[1] * $M[2][1]);
    $y = ($v[0] * $M[1][2]) + ($v[1] * $M[2][2]);

    $v[0] = $x;
    $v[1] = $y;

    echo "Day(" . $i . ")=[" . $v[0] . " - " . $v[1] .
"]</br>";
}

Function ExtractProb($s) {
global $M;

$Eb = "S";
$Es = "R";

for($i=1; $i<=2; $i++){
    for($j=1; $j<=2; $j++){
        $M[$i][$j]=0;
    }
}

$TB = 0;
$TS = 0;

for($i=1; $i<strlen($s)-1; $i++){

        $DI1 = substr($s, $i, 1);
        $DI2 = substr($s, $i+1, 1);

        If ($DI1 == $Eb) {$r = 1;}
        If ($DI1 == $Es) {$r = 2;}
        If ($DI2 == $Eb) {$c = 1;}
        If ($DI2 == $Es) {$c = 2;}

        $M[$r][$c] = $M[$r][$c] + 1;
```

```
            If ($DI1 == $Eb) {$TB = $TB + 1;}
            If ($DI1 == $Es) {$TS = $TS + 1;}
}

for($i=1; $i<=2; $i++){
    for($j=1; $j<=2; $j++){
        If ($i == 1) {$M[$i][$j] = intval($M[$i][$j]) / $TB;}
        If ($i == 2) {$M[$i][$j] = intval($M[$i][$j]) / $TS;}
    }
}

}
?>
```

```
PHP Output:
Day(1)=[0.2 - 0.8]
Day(2)=[0.54 - 0.46]
Day(3)=[0.3955 - 0.6045]
Day(4)=[0.4569125 - 0.5430875]
Day(5)=[0.4308121875 - 0.5691878125]
```

Supporting algorithm (PHP alternative) 8.

```
<?php
$M = array();
$v = array();

ExtractProb("SRCCRRSSCSRCSR");

$chain = 5;

$v[0] = 0;
$v[1] = 1;
$v[2] = 0;

for($i=1; $i<=$chain; $i++){
    $x = ($v[0] * $M[1][1]) + ($v[1] * $M[2][1]) + ($v[2] *
    $M[3][1]);
    $y = ($v[0] * $M[1][2]) + ($v[1] * $M[2][2]) + ($v[2] *
    $M[3][2]);
    $z = ($v[0] * $M[1][3]) + ($v[1] * $M[2][3]) + ($v[2] *
    $M[3][3]);
```

```php
    $v[0] = $x;
    $v[1] = $y;
    $v[2] = $z;

echo "Day(" . $i . ")=[" . $v[0] . "|" . $v[1] . "|"
. $v[2] . "]</br>";
}

Function ExtractProb($s) {
global $M;

$Eb = "S";
$Es = "R";
$Ec = "C";

for($i=1; $i<=3; $i++){
    for($j=1; $j<=3; $j++){
        $M[$i][$j]=0;
    }
}

$TB = 0;
$TS = 0;
$TC = 0;

for($i=1; $i<strlen($s)-1; $i++){

        $DI1 = substr($s, $i, 1);
        $DI2 = substr($s, $i+1, 1);

        If ($DI1 == $Eb) {$r = 1;}
        If ($DI1 == $Es) {$r = 2;}
        If ($DI1 == $Ec) {$r = 3;}
        If ($DI2 == $Eb) {$c = 1;}
        If ($DI2 == $Es) {$c = 2;}
        If ($DI2 == $Ec) {$c = 3;}

        $M[$r][$c] = $M[$r][$c] + 1;

        If ($DI1 == $Eb) {$TB = $TB + 1;}
        If ($DI1 == $Es) {$TS = $TS + 1;}
        If ($DI1 == $Ec) {$TC = $TC + 1;}
```

```
}
for($i=1; $i<=3; $i++){
    for($j=1; $j<=3; $j++){
        If ($i == 1) {$M[$i][$j] = intval($M[$i][$j]) / $TB;}
        If ($i == 2) {$M[$i][$j] = intval($M[$i][$j]) / $TS;}
        If ($i == 3) {$M[$i][$j] = intval($M[$i][$j]) / $TC;}
    }
}
}
?>
```

```
PHP Output:
Day(1) = [0.25|0.25|0.5]
Day(2) = [0.375|0.3125|0.3125]
Day(3) = [0.328125|0.34375|0.328125]
Day(4) = [0.33203125|0.33203125|0.3359375]
Day(5) = [0.333984375|0.3330078125|0.3330078125]
```

Supporting algorithm (PHP alternative) 9.

```php
<?php
$M = array();
$v = array();

ExtractProb("TACTTCGATTTAAGCGCGGCGGCCTATATTA");

$chain =3;

$v[0] = 1;
$v[1] = 0;
$v[2] = 0;
$v[3] = 0;

for($i=1; $i<=$chain; $i++){

$x=($v[0]*$M[1][1])+($v[1]*$M[2][1])+($v[2]*$M[3][1])+($v[3]*
$M[4][1]);
$y=($v[0]*$M[1][2])+($v[1]*$M[2][2])+($v[2]*$M[3][2])+($v[3]*
$M[4][2]);
$z=($v[0]*$M[1][3])+($v[1]*$M[2][3])+($v[2]*$M[3][3])+($v[3]*
$M[4][3]);
$w=($v[0]*$M[1][4])+($v[1]*$M[2][4])+($v[2]*$M[3][4])+($v[3]*
$M[4][4]);
```

```
      $v[0] = $x;
      $v[1] = $y;
      $v[2] = $z;
      $v[3] = $w;

echo "Base(".$i.")=[".$v[0]."|".$v[1]."|".$v[2]."|".$v[3].
  "]</br>";
$Baseby = $Baseby . Base($v);
}

echo "BasesPredicted[" . $Baseby . "]";

Function Base($v){
  for($i=0; $i<=count($v); $i++){
     If ($v[$i] > $old){
          $x = $v[$i];
          $h = $i;
     }
     $old = $x;
  }
     If ($h == 0) {$n = "A";}
     If ($h == 1) {$n = "T";}
     If ($h == 2) {$n = "G";}
     If ($h == 3) {$n = "C";}
     Return $n;
}

Function ExtractProb($s) {
global $M;

$Ea = "A";
$Et = "T";
$Eg = "G";
$Ec = "C";

for($i=1; $i<=4; $i++){
    for($j=1; $j<=4; $j++){
        $M[$i][$j]=0;
    }
}

$Ta = 0;
$Tt = 0;
$Tg = 0;
```

```php
$Tc = 0;

for($i=1; $i<strlen($s)-1; $i++){

        $DI1 = substr($s, $i, 1);
        $DI2 = substr($s, $i+1, 1);

        If ($DI1 == $Ea) {$r = 1;}
        If ($DI1 == $Et) {$r = 2;}
        If ($DI1 == $Eg) {$r = 3;}
        If ($DI1 == $Ec) {$r = 4;}

        If ($DI2 == $Ea) {$c = 1;}
        If ($DI2 == $Et) {$c = 2;}
        If ($DI2 == $Eg) {$c = 3;}
        If ($DI2 == $Ec) {$c = 4;}

        $M[$r][$c] = $M[$r][$c] + 1;

        If ($DI1 == $Ea) {$Ta = $Ta + 1;}
        If ($DI1 == $Et) {$Tt = $Tt + 1;}
        If ($DI1 == $Eg) {$Tg = $Tg + 1;}
        If ($DI1 == $Ec) {$Tc = $Tc + 1;}
}

for($i=1; $i<=4; $i++){
    for($j=1; $j<=4; $j++){
        If ($i == 1) {$M[$i][$j] = intval($M[$i][$j]) / $Ta;}
        If ($i == 2) {$M[$i][$j] = intval($M[$i][$j]) / $Tt;}
        If ($i == 3) {$M[$i][$j] = intval($M[$i][$j]) / $Tg;}
        If ($i == 4) {$M[$i][$j] = intval($M[$i][$j]) / $Tc;}
    }
}
}
?>
```

```
PHP Output:
Base(1)=[0.16666666666|0.50000000000000|
         0.16666666666667|0.16666666666667]
Base(2)=[0.27380952380|0.35317460317460|
         0.17063492063492|0.20238095238095]
Base(3)=[0.22697782816|0.35169438145629|
         0.21003401360544|0.21129377676997]
BasesPredicted[TTT]
```

Supporting algorithm (PHP alternative) 10.

```php
<?php
$M = array();
$v = array();

ExtractProb("TACTTCGATTTAAGCGCGGCGGCCTATATTA");

$chain =5;

$v[0][0] = 1;
$v[1][0] = 0;
$v[2][0] = 0;
$v[3][0] = 0;

$v[0][1] = 0;
$v[1][1] = 0;
$v[2][1] = 0;
$v[3][1] = 0;

for($k=1; $k<=$chain; $k++){
    for($i=0; $i<=3; $i++){
        for($j=0; $j<=3; $j++){
            $v[$i][1] = $v[$i][1] + ($v[$j][0] * $M[$j + 1]
            [$i + 1]);
        }
    }

for($i=0; $i<=3; $i++){
    $v[$i][0] = $v[$i][1];
    $v[$i][1] = 0;
}

    $A = $v[0][0];
    $T = $v[1][0];
    $G = $v[2][0];
    $C = $v[3][0];

echo "V(".$k.")=[".$A."|".$T."|".$G."|".$C."]</br>";
}

Function ExtractProb($s) {
global $M;
```

```php
$Ea = "A";
$Et = "T";
$Eg = "G";
$Ec = "C";

for($i=1; $i<=4; $i++){
    for($j=1; $j<=4; $j++){
        $M[$i][$j]=0;
    }
}

$Ta = 0;
$Tt = 0;
$Tg = 0;
$Tc = 0;

for($i=1; $i<strlen($s)-1; $i++){

        $DI1 = substr($s, $i, 1);
        $DI2 = substr($s, $i+1, 1);

        If ($DI1 == $Ea) {$r = 1;}
        If ($DI1 == $Et) {$r = 2;}
        If ($DI1 == $Eg) {$r = 3;}
        If ($DI1 == $Ec) {$r = 4;}

        If ($DI2 == $Ea) {$c = 1;}
        If ($DI2 == $Et) {$c = 2;}
        If ($DI2 == $Eg) {$c = 3;}
        If ($DI2 == $Ec) {$c = 4;}

        $M[$r][$c] = $M[$r][$c] + 1;

        If ($DI1 == $Ea) {$Ta = $Ta + 1;}
        If ($DI1 == $Et) {$Tt = $Tt + 1;}
        If ($DI1 == $Eg) {$Tg = $Tg + 1;}
        If ($DI1 == $Ec) {$Tc = $Tc + 1;}
}

for($i=1; $i<=4; $i++){
    for($j=1; $j<=4; $j++){
        If ($i == 1) {$M[$i][$j] = intval($M[$i][$j]) / $Ta;}
        If ($i == 2) {$M[$i][$j] = intval($M[$i][$j]) / $Tt;}
```

```
        If ($i == 3) {$M[$i][$j] = intval($M[$i][$j]) / $Tg;}
        If ($i == 4) {$M[$i][$j] = intval($M[$i][$j]) / $Tc;}
    }
}
}
?>
```

```
PHP Output:
V(1)=[0.16666666666666|0.50000000000000|
      0.16666666666666|0.16666666666666]
V(2)=[0.27380952380952|0.35317460317460|
      0.17063492063492|0.20238095238095]
V(3)=[0.22697782816830|0.35169438145628|
      0.21003401360544|0.21129377676996]
V(4)=[0.22414311109511|0.33016717857042|
      0.21857865721244|0.22711105312201]
V(5)=[0.21532367589496|0.32370139612165|
      0.22958597474151|0.2313889532418]
```

Supporting algorithm (PHP alternative) 11.

```php
<?php
$Inp = array();
$Obs = "";
$Reg = "";
$l="";

$R = "159";

$Inp = explode(",",$R);
$Lu = 200;
$Ld = 60;
$n = 4;

$Pr = ($Lu - $Ld) / $n;

for($i=0; $i<count($Inp); $i++){

    $s = ($Inp[$i] - $Ld) / $Pr;
    $s = floor($s);

    if($s == 0){$l = "A";}
    if($s == 1){$l = "B";}
```

```php
    if($s == 2){$l = "C";}
    if($s == 3){$l = "D";}

    $Obs = $Obs . $l;
    $Reg = $Reg . $s . ",";
}

echo "Reg=" . $Reg . "</br>" . "Obs=" . $Obs . "</br>";

?>
```

```
PHP Output:
Reg=2,0,3,3,3,1,3,1,1,1,1,1,0,2,2,1,1,0,0,3,3,3,2,0,2,1,3,3,0,3,
    0,3,3,3,2,3,1,1,3,1,1,3,
Obs=CADDDBDBBBBBBACCBBAADDDCACBDDADADDDCDBBDBBD
```

Supporting algorithm (PHP alternative) 12.

```php
<?php
$Jar = array();
$v = array();

$Jar[1][1] = 0.33;
$Jar[1][2] = 0.33;
$Jar[1][3] = 0.33;

$Jar[2][1] = 0.5;
$Jar[2][2] = 0.5;
$Jar[2][3] = 0;

$Jar[3][1] = 0;
$Jar[3][2] = 0;
$Jar[3][3] = 1;

$chain =5;

$v[0][0] = 1;
$v[1][0] = 0;
$v[2][0] = 0;

$v[0][1] = 0;
$v[1][1] = 0;
$v[2][1] = 0;
```

```php
for($k=1; $k<=$chain; $k++){

    for($i=0; $i<=2; $i++){
        for ($j=0; $j<=2; $j++){
            $v[$i][1] = $v[$i][1] + ($v[$j][0] * $Jar[$j + 1]
            [$i + 1]);
        }
    }

for($i=0; $i<=2; $i++){
    $v[$i][0] = $v[$i][1];
    $v[$i][1] = 0;
}
    $A = $v[0][0];
    $B = $v[1][0];
    $C = $v[2][0];

echo "Step(".$k.")=[".$A."|".$B."|".$C."]</br>";
}
?>
```

```
PHP Output:
Step(1)=[0.33|0.33|0.33]
Step(2)=[0.2739|0.2739|0.4389]
Step(3)=[0.227337|0.227337|0.529287]
Step(4)=[0.18868971|0.18868971|0.60430821]
Step(5)=[0.1566124593|0.1566124593|0.6665758143]
```

Supporting algorithm (PHP alternative) 13.

```php
<?php
$Jar = array();
$v = array();

$Jar[1][1] = 1;
$Jar[1][2] = 0;
$Jar[1][3] = 0;
$Jar[1][4] = 0;

$Jar[2][1] = 0.5;
$Jar[2][2] = 0;
$Jar[2][3] = 0.5;
$Jar[2][4] = 0;
```

```php
$Jar[3][1] = 0;
$Jar[3][2] = 0.5;
$Jar[3][3] = 0;
$Jar[3][4] = 0.5;

$Jar[4][1] = 0;
$Jar[4][2] = 0;
$Jar[4][3] = 1;
$Jar[4][4] = 0;

$chain = 5;

$v[0][0] = 0;
$v[1][0] = 0;
$v[2][0] = 0;
$v[3][0] = 1;

$v[0][1] = 0;
$v[1][1] = 0;
$v[2][1] = 0;
$v[3][1] = 0;

for($k=1; $k<=$chain; $k++){

    for($i=0; $i<=3; $i++){
        for($j=0; $j<=3; $j++){
            $v[$i][1] = $v[$i][1] + ($v[$j][0] * $Jar[$j + 1]
            [$i + 1]);
        }
    }

for($i=0; $i<=3; $i++){
    $v[$i][0] = $v[$i][1];
    $v[$i][1] = 0;
}
    $A = $v[0][0];
    $B = $v[1][0];
    $C = $v[2][0];
    $D = $v[3][0];

echo "Step(".$k.")=[".$A."|".$B."|".$C."|".$D."]</br>";
}
?>
```

```
PHP Output:
Step(1) = [0|0|1|0]
Step(2) = [0|0.5|0|0.5]
Step(3) = [0.25|0|0.75|0]
Step(4) = [0.25|0.375|0|0.375]
Step(5) = [0.4375|0|0.5625|0]
```

Supporting algorithm (PHP alternative) 14.

```php
<?php
$P = array();
$Jar = array();

$P[0][0] = "A";
$P[0][1] = "B";
$P[0][2] = "C";

$P[1][0] = 0.33;
$P[1][1] = 0.33;
$P[1][2] = 0.33;

$P[2][0] = 0;
$P[2][1] = 0.5;
$P[2][2] = 0.5;

$P[3][0] = 1;
$P[3][1] = 0;
$P[3][2] = 0;

for($j=0; $j<=3; $j++){
    $Jar[$j] = Fill_Jar($j);
    if($j>0){echo "Jar(".$j.")=(".$Jar[$j].")</br>";}
}

$draws = 20;
$a = Draw(1);

for ($i=1; $i<=$draws; $i++){
    for ($j=0; $j<=3; $j++){

        If ($a == $P[0][$j]) {
            $a = Draw($j + 1);
            $q = $q . $P[0][$j];
```

```php
                $z = $z . ", Jar " . $P[0][$j] . "[" . $a . "]";
                $j=3;
            }
        }
}

echo "Q = " . $q . "</br>";
echo "Z = " . $z;

function Draw($S) {
    srand(mktime());
    global $Jar;
    $randomly_choose = rand(1,strlen($Jar[$S])-1);
    $ball = substr($Jar[$S], $randomly_choose, 1);
    return $ball;
}

function Fill_Jar($S){
global $P;

$Ltot = 27;
$b = "";
    for ($i=0; $i<=2; $i++){
    $a = round($Ltot * $P[$S][$i]);
        for ($j=1; $j<=$a; $j++){
        $b = $b . $P[0][$i];
        }
    }
return $b;
}
?>
```

```
PHP Output:
Jar(1) = (AAAAAAAAABBBBBBBBBCCCCCCCCC)
Jar(2) = (BBBBBBBBBBBBBBBCCCCCCCCCCCCCC)
Jar(3) = (AAAAAAAAAAAAAAAAAAAAAAAAAAA)

Q = CACABCACABBBBBCACAAB

Z ="Jar C[A], Jar A[C], Jar C[A], Jar A[B], Jar B[C],
    Jar C[A], Jar A[C], Jar C[A], Jar A[B], Jar B[B],
    Jar B[B], Jar B[B], Jar B[B], Jar B[C], Jar C[A],
    Jar A[C], Jar C[A], Jar A[A], Jar A[B], Jar B[B]"
```

Supporting algorithm (PHP alternative) 15.

```php
<?php
$P = array();
$Jar = array();

$P[0][0] = "A";
$P[0][1] = "B";
$P[0][2] = "C";
$P[0][3] = "D";

$P[1][0] = 0;
$P[1][1] = 1;
$P[1][2] = 0;
$P[1][3] = 0;

$P[2][0] = 0.33;
$P[2][1] = 0;
$P[2][2] = 0.33;
$P[2][3] = 0.33;

$P[3][0] = 0;
$P[3][1] = 1;
$P[3][2] = 0;
$P[3][3] = 0;

$P[4][0] = 0;
$P[4][1] = 0;
$P[4][2] = 1;
$P[4][3] = 0;

for($j=1; $j<=4; $j++){
    $Jar[$j] = Fill_Jar($j);
}

$draws =100;
$a = Draw(1);

for ($i=1; $i<=$draws; $i++){
    for ($j=0; $j<=3; $j++){
        If ($a == $P[0][$j]){
            $a = Draw($j + 1);
```

```php
                $q = $q . $P[0][$j];
                $j=3;
            }
        }
}

echo "Q = " . $q . "</br>";

function Draw($S) {
    srand(mktime());
    global $Jar;
    $randomly_choose = rand(1,strlen($Jar[$S])-1);
    $ball = substr($Jar[$S], $randomly_choose, 1);
    return $ball;
}

function Fill_Jar($S){
global $P;

$Ltot = 100;
$b = "";
    for ($i=0; $i<=3; $i++){
    $a = round($Ltot * $P[$S][$i]);
        for ($j=1; $j<=$a; $j++){
            $b = $b . $P[0][$i];
        }
    }
return $b;
}
?>
```

```
PHP Output:
Q = BABDCBDCBABDCBCBDCBABABABDCBDCBDCBCBABABDCBDCBDCBA
    BCBABCBDCBDCBDCBDCBABCBCBABABDCBDCBCBDCBCBDCBDCBAB
```

Supporting algorithm (PHP alternative) 16.

```php
<?php
$P = array();

ExtractProb("BABDCBDCBABDCBCBDCBABABABDCBDCBDCBCBABABD
CBDCBDCBABCBABCBDCBDCBDCBDCBABCBCBABABDCBDCBCBDCBCBDCBDCBAB");
```

```php
$z = "<table border='1'>";
for($i=1; $i<=4; $i++){
$z .= "<tr>";
    for($j=1; $j<=4; $j++){
        $z .= "<td>" . Round($P[$i][$j], 2) . "</td>" ;
    }
$z .= "</tr>";
}
$z .= "</table>";

echo $z;

Function ExtractProb($s) {
global $P;

$Ea = "A";
$Eb = "B";
$Ec = "C";
$Ed = "D";

for($i=1; $i<=4; $i++){
    for($j=1; $j<=4; $j++){
        $P[$i][$j]=0;
    }
}

$Ta = 0;
$Tb = 0;
$Tc = 0;
$Td = 0;

for($i=1; $i<strlen($s)-1; $i++){

        $DI1 = substr($s, $i, 1);
        $DI2 = substr($s, $i+1, 1);

        If ($DI1 == $Ea) {$r = 1;}
        If ($DI1 == $Eb) {$r = 2;}
        If ($DI1 == $Ec) {$r = 3;}
        If ($DI1 == $Ed) {$r = 4;}
        If ($DI2 == $Ea) {$c = 1;}
        If ($DI2 == $Eb) {$c = 2;}
```

```
        If ($DI2 == $Ec) {$c = 3;}
        If ($DI2 == $Ed) {$c = 4;}

        $P[$r][$c] = $P[$r][$c] + 1;

        If ($DI1 == $Ea) {$Ta = $Ta + 1;}
        If ($DI1 == $Eb) {$Tb = $Tb + 1;}
        If ($DI1 == $Ec) {$Tc = $Tc + 1;}
        If ($DI1 == $Ed) {$Td = $Td + 1;}
}

for($i=1; $i<=4; $i++){
    for($j=1; $j<=4; $j++){
        If ($i == 1) {$P[$i][$j] = intval($P[$i][$j]) / $Ta;}
        If ($i == 2) {$P[$i][$j] = intval($P[$i][$j]) / $Tb;}
        If ($i == 3) {$P[$i][$j] = intval($P[$i][$j]) / $Tc;}
        If ($i == 4) {$P[$i][$j] = intval($P[$i][$j]) / $Td;}
    }
}
}
?>
```

```
PHP Output:
```

0	1	0	0
0.31	0	0.21	0.49
0	1	0	0
0	0	1	0

Supporting algorithm (PHP alternative) 17.

```php
<?php
$P = array();
$Jar = array();
$f = array();

$P[0][0] = "A";
$P[0][1] = "B";
$P[0][2] = "C";
$P[0][3] = "D";

$P[1][0] = 0;
$P[1][1] = 1;
```

```
$P[1][2] = 0;
$P[1][3] = 0;

$P[2][0] = 0.33;
$P[2][1] = 0;
$P[2][2] = 0.33;
$P[2][3] = 0.33;

$P[3][0] = 0;
$P[3][1] = 1;
$P[3][2] = 0;
$P[3][3] = 0;

$P[4][0] = 0;
$P[4][1] = 0;
$P[4][2] = 1;
$P[4][3] = 0;

for($j=1; $j<=4; $j++){
    $Jar[$j] = Fill_Jar($j);
}

$draws = 10000;
$a = Draw(2);

for ($i=1; $i<=$draws; $i++){
    for ($j=0; $j<=3; $j++){
        If ($a == $P[0][$j]){
            $a = Draw($j + 1);
            $z .= $a;
            $j=3;
        }
    }
}

for($i=1; $i<strlen($z); $i++){
    $g = substr($z, $i, 1);
    If ($g == "A"){$f[0] = $f[0] + 1;}
    If ($g == "B"){$f[1] = $f[1] + 1;}
    If ($g == "C"){$f[2] = $f[2] + 1;}
    If ($g == "D"){$f[3] = $f[3] + 1;}
}
```

```php
for($i=0; $i<=3; $i++){
$pro .= $P[0][$i] . "=" . round((100 / strlen($z)) * $f[$i]) .
"% ";
}

echo $pro;

function Draw($S) {
    srand(mktime());
    global $Jar;
    $randomly_choose = rand(1,strlen($Jar[$S])-1);
    $ball = substr($Jar[$S], $randomly_choose, 1);
    return $ball;
}

function Fill_Jar($S){
global $P;

$Ltot = 100;
$b = "";
    for ($i=0; $i<=3; $i++){
    $a = round($Ltot * $P[$S][$i]);
        for ($j=1; $j<=$a; $j++){
            $b = $b . $P[0][$i];
        }
    }
return $b;
}
?>
```

```
PHP Output:
A=14%  B=43%  C=29%  D=14%
```

B

Supporting Algorithms in JavaScript

Supporting algorithm (JS alternative) 1.

```
<script>
var Jar = [];

Jar[0] = "WWBBBBBBBB";
Jar[1] = "WWWWWBBBBB";

var draws = 17;
var z;

a = Draw(1);
z = z + " Jar W[" + a + "],";

for (var i=1; i<=draws; i++){
    if (a === "W") {
        a = Draw(0);
        z = z + " Jar W[" + a + "],";
    } else {
        a = Draw(1);
        z = z + " Jar B[" + a + "],";
    }
}

document.write(z);

function Draw(S) {
    randomly_choose = Math.floor((Math.random() *
                        Jar[S].length));
    ball = Jar[S].substr(randomly_choose, 1);
    return ball;
}
</script>
```

Markov Chains: From Theory to Implementation and Experimentation, First Edition. Paul A. Gagniuc.
© 2017 John Wiley & Sons, Inc. Published 2017 by John Wiley & Sons, Inc.
Companion website: www.wiley.com/go/gagniuc/markovchains

```
JavaScript Output:
Jar W[B], Jar B[W], Jar W[W], Jar W[W], Jar W[B], Jar B[W],
Jar W[B], Jar B[W], Jar W[B], Jar B[B], Jar B[W], Jar W[B],
Jar B[W], Jar W[B], Jar B[B], Jar B[B], Jar B[W], Jar W[B],
```

Supporting algorithm (JS alternative) 2.

```
<script>
var draws = 17;
var z="";

var Jar = [];

Jar[0]="";
Jar[1]="";

Fill_Jar(0, 0.2);
Fill_Jar(1, 0.6);

a = Draw(1);
z = z + " Jar W[" + a + "],";

for (var i=1; i<=draws; i++){
    if (a === "W") {
        a = Draw(0);
        z = z + " Jar W[" + a + "],";
    } else {
        a = Draw(1);
        z = z + " Jar B[" + a + "],";
    }
}

document.write(z);

function Draw(S) {
    randomly_choose = Math.floor((Math.random() *
                        Jar[S].length));
    ball = Jar[S].substr(randomly_choose, 1);
    return ball;
}
```

```
function Fill_Jar(S, p){
    var Balls_W = Math.round(100 * p);
    var Balls_B = 100 - Balls_W;

    for (var i=1; i<=Balls_W; i++){
        Jar[S] = Jar[S] + "W";
    }
    for (var i=1; i<=Balls_B; i++){
        Jar[S] = Jar[S] + "B";
    }
}
</script>
```

```
JavaScript Output:
Jar W[B], Jar B[W], Jar W[B], Jar B[B], Jar B[W], Jar W[B],
Jar B[W], Jar W[B], Jar B[W], Jar W[B], Jar B[W], Jar W[B],
Jar B[W], Jar W[W], Jar W[B], Jar B[W], Jar W[B], Jar B[B],
```

Supporting algorithm (JS alternative) 3.

```
<script>
var M = [];

ExtractProb("SRRSRSRRSRSRRSS");

function ExtractProb(s) {
var Eb = "S";
var Es = "R";

for(var i=1; i<=2; i++){
    M[i]=[];
    for(var j=1; j<=2; j++){
        M[i][j]=0;
    }
}

var TB = 0;
var TS = 0;
var r;
var c;
```

```
for(var i=1; i<s.length-1; i++){

        var DI1 = s.substr(i, 1);
        var DI2 = s.substr(i+1, 1);

        if (DI1 === Eb) {r = 1;}
        if (DI1 === Es) {r = 2;}
        if (DI2 === Eb) {c = 1;}
        if (DI2 === Es) {c = 2;}

        M[r][c] = M[r][c] + 1;

        if (DI1 === Eb) {TB = TB + 1;}
        if (DI1 === Es) {TS = TS + 1;}
}

DrowMatrix(2, 2, "(C)", "Count:");

for(var i=1; i<=2; i++){
    for(var j=1; j<=2; j++){
        if (i === 1) {M[i][j] = Math.round(M[i][j]) / TB;}
        if (i === 2) {M[i][j] = Math.round(M[i][j]) / TS;}
    }
}

DrowMatrix(2, 2, "(P)", "Transition matrix M:");
}

function DrowMatrix(ib, jb, model, msg) {
ct = "";
Eb = "S";
Es = "R";

ct += "<table border='1'><tr>";
ct += "<td>" + model + "</td><td>" + Eb + "</td><td>" + Es +
"</td></tr>";

for(var i=1; i<=ib; i++){
        ct += "<tr>";
        for(var j=1; j<=jb; j++){
                v = M[i][j].toFixed(1);
                if (j === 1 && i === 1) {ct += "<td>" + Eb +
"</td>";}
```

```
                    if (j === 1 && i === 2) {ct += "<td>" + Es +
"</td>";}
                ct += "<td>" + v + "</td>";
        }
ct += "</tr>";
}

ct += "</table>";
rez = msg + " M[" + jb + "," + ib + "]" + "</br>" + ct +
"</br>";

document.write(rez);
}
</script>
```

JavaScript Output:		
Count: M[2,2]		
(C)	S	R
S	1.0	4.0
R	5.0	3.0
Transition matrix M: M[2,2]		
(P)	S	R
S	0.2	0.8
R	0.6	0.4

Supporting algorithm (JS alternative) 4.

```
<script>
var P = [];
var v = [];

var chain = 5;

for(var i=1; i<=2; i++){
    P[i]=[];
}

P[1][1] = 0.2;
P[1][2] = 0.625;
P[2][1] = 0.8;
P[2][2] = 0.375;
```

```
v[0] = 1;
v[1] = 0;

for(var i=1; i<=chain; i++){
    x = (v[0] * P[1][1]) + (v[1] * P[1][2]);
    y = (v[0] * P[2][1]) + (v[1] * P[2][2]);

    v[0] = x;
    v[1] = y;

    document.write(v[0] + " | " + v[1] + "</br>");
}
</script>
```

```
JavaScript Output:
0.2 | 0.8
0.54 | 0.4600000000000001
0.3955000000000001 | 0.6045
0.4569125 | 0.5430875000000001
0.43081218750000005 | 0.5691878125000001
```

Supporting algorithm (JS alternative) 5.

```
<script>
var P = [];
var v = [];

var chain = 50;

for(var i=1; i<=2; i++){
    P[i]=[];
}

P[1][1] = 0.2;
P[1][2] = 0.625;
P[2][1] = 0.8;
P[2][2] = 0.375;

v[0] = 1;
v[1] = 0;
```

```
for(var i=1; i<=chain; i++){
    x = (v[0] * P[1][1]) + (v[1] * P[1][2]);
    y = (v[0] * P[2][1]) + (v[1] * P[2][2]);

    if(v[0] === x && v[1] === y){
        document.write("Steady state vector at day [" + i +
"]!");
        i = chain;
    } else {
        document.write("Day[" + i + "], v=[" + x + " | " + y
+ "] </br>");
    }

    v[0] = x;
    v[1] = y;
}
</script>
```

Supporting algorithm (JS alternative) 6.

```
<script>
var P = [];
var v = [];

var chain = 5;

for(var i=1; i<=2; i++){
    P[i]=[];
}

P[1][1] = 0.2;
P[1][2] = 0.625;
P[2][1] = 0.8;
P[2][2] = 0.375;

a = P[1][1];
b = P[2][2];

x = (1 - b) / (2 - (a + b));
y = (1 - a) / (2 - (a + b));
```

```
v[0] = x;
v[1] = y;

document.write("Steady state vector v = [" + v[0] + " | " +
v[1] + "]");
</script>
```

> **JavaScript Output:**
> **Steady state vector v = [0.43859649122807 | 0.56140350877193]**

Supporting algorithm (JS alternative) 7.

```
<script>
var M = [];
var v = [];

ExtractProb("SRRSRSRRSRSRRSS");

var chain = 5;

v[0] = 1;
v[1] = 0;

for(var i=1; i<=chain; i++){
    x = (v[0] * M[1][1]) + (v[1] * M[2][1]);
    y = (v[0] * M[1][2]) + (v[1] * M[2][2]);

    v[0] = x;
    v[1] = y;

    document.write("Day(" + i + ")=[" + v[0] + " | " + v[1]
+ "]</br>");
}

function ExtractProb(s) {
var Eb = "S";
var Es = "R";

for(var i=1; i<=2; i++){
    M[i]=[];
```

```
    for(var j=1; j<=2; j++){
        M[i][j]=0;
    }
}

var TB = 0;
var TS = 0;
var r;
var c;

for(var i=1; i<s.length-1; i++){

        var DI1 = s.substr(i, 1);
        var DI2 = s.substr(i+1, 1);

        if (DI1 === Eb) {r = 1;}
        if (DI1 === Es) {r = 2;}
        if (DI2 === Eb) {c = 1;}
        if (DI2 === Es) {c = 2;}

        M[r][c] = M[r][c] + 1;

        if (DI1 === Eb) {TB = TB + 1;}
        if (DI1 === Es) {TS = TS + 1;}
}

for(var i=1; i<=2; i++){
    for(var j=1; j<=2; j++){
        if (i === 1) {M[i][j] = Math.round(M[i][j]) / TB;}
        if (i === 2) {M[i][j] = Math.round(M[i][j]) / TS;}
    }
}
}
</script>
```

```
JavaScript Output:
Day(1)=[0.2 - 0.8]
Day(2)=[0.54 - 0.46]
Day(3)=[0.3955 - 0.6045]
Day(4)=[0.4569125 - 0.5430875]
Day(5)=[0.4308121875 - 0.5691878125]
```

Supporting algorithm (JS alternative) 8.

```
<script>
var M = [];
var v = [];

ExtractProb("SRCCRRSSCSRCSR");

var chain = 5;

v[0] = 0;
v[1] = 1;
v[2] = 0;

for(var i=1; i<=chain; i++){
    x = (v[0] * M[1][1]) + (v[1] * M[2][1]) + (v[2] * M[3][1]);
    y = (v[0] * M[1][2]) + (v[1] * M[2][2]) + (v[2] * M[3][2]);
    z = (v[0] * M[1][3]) + (v[1] * M[2][3]) + (v[2] * M[3][3]);

  v[0] = x;
  v[1] = y;
  v[2] = z;

document.write("Day("+i+")=["+v[0]+"|"+v[1]+"|"+v[2]+
"]</br>");
}

function ExtractProb(s) {
var Eb = "S";
var Es = "R";
var Ec = "C";

for(var i=1; i<=3; i++){
    M[i]=[];
    for(var j=1; j<=3; j++){
        M[i][j]=0;
    }
}

var TB = 0;
var TS = 0;
```

```
var TC = 0;
var r;
var c;

for(var i=1; i<s.length-1; i++){

        var DI1 = s.substr(i, 1);
        var DI2 = s.substr(i+1, 1);

        if (DI1 === Eb) {r = 1;}
        if (DI1 === Es) {r = 2;}
        if (DI1 === Ec) {r = 3;}
        if (DI2 === Eb) {c = 1;}
        if (DI2 === Es) {c = 2;}
        if (DI2 === Ec) {c = 3;}

        M[r][c] = M[r][c] + 1;

        if (DI1 === Eb) {TB = TB + 1;}
        if (DI1 === Es) {TS = TS + 1;}
        if (DI1 === Ec) {TC = TC + 1;}
}

for(var i=1; i<=3; i++){
    for(var j=1; j<=3; j++){
        if (i === 1) {M[i][j] = Math.round(M[i][j]) / TB;}
        if (i === 2) {M[i][j] = Math.round(M[i][j]) / TS;}
        if (i === 3) {M[i][j] = Math.round(M[i][j]) / TC;}
    }
}
}
</script>
```

```
JavaScript Output:
Day(1) = [0.25|0.25|0.5]
Day(2) = [0.375|0.3125|0.3125]
Day(3) = [0.328125|0.34375|0.328125]
Day(4) = [0.33203125|0.33203125|0.3359375]
Day(5) = [0.333984375|0.3330078125|0.3330078125]
```

Supporting algorithm (JS alternative) 9.

```
<script>
var M = [];
var v = [];

ExtractProb("TACTTCGATTTAAGCGCGGCGGCCTATATTA");

var chain = 3;
var Baseby = "";

v[0] = 1;
v[1] = 0;
v[2] = 0;
v[3] = 0;

for(var i=1; i<=chain; i++){
x = (v[0]*M[1][1])+(v[1]*M[2][1])+(v[2]*M[3][1])+(v[3]*M[4][1]);
y = (v[0]*M[1][2])+(v[1]*M[2][2])+(v[2]*M[3][2])+(v[3]*M[4][2]);
z = (v[0]*M[1][3])+(v[1]*M[2][3])+(v[2]*M[3][3])+(v[3]*M[4][3]);
w = (v[0]*M[1][4])+(v[1]*M[2][4])+(v[2]*M[3][4])+(v[3]*M[4][4]);

    v[0] = x;
    v[1] = y;
    v[2] = z;
    v[3] = w;

document.write("Day("+i+") = ["+v[0]+"|"+v[1]+"|"+v[2]+"|"+v[3]
+"]</br>");
Baseby = Baseby + Base(v);
}

document.write("BasesPredicted[" + Baseby + "]");

function Base(v){
var x, h;
var old=0;

  for(var i=0; i<=v.length; i++){
     if(v[i] > old){
         x = v[i];
         h = i;
     }
```

```
      old = x;
  }
    if(h === 0) {n = "A";}
    if(h === 1) {n = "T";}
    if(h === 2) {n = "G";}
    if(h === 3) {n = "C";}
    return n;
}

function ExtractProb(s) {
var Ea = "A";
var Et = "T";
var Eg = "G";
var Ec = "C";

for(var i=1; i<=4; i++){
    M[i]=[];
    for(var j=1; j<=4; j++){
        M[i][j]=0;
    }
}

var Ta = 0;
var Tt = 0;
var Tg = 0;
var Tc = 0;
var r;
var c;

for(var i=1; i<s.length-1; i++){

        var DI1 = s.substr(i, 1);
        var DI2 = s.substr(i+1, 1);

        if (DI1 === Ea) {r = 1;}
        if (DI1 === Et) {r = 2;}
        if (DI1 === Eg) {r = 3;}
        if (DI1 === Ec) {r = 4;}

        if (DI2 === Ea) {c = 1;}
        if (DI2 === Et) {c = 2;}
        if (DI2 === Eg) {c = 3;}
        if (DI2 === Ec) {c = 4;}
```

```
            M[r][c] = M[r][c] + 1;

            if (DI1 === Ea) {Ta = Ta + 1;}
            if (DI1 === Et) {Tt = Tt + 1;}
            if (DI1 === Eg) {Tg = Tg + 1;}
            if (DI1 === Ec) {Tc = Tc + 1;}
}

for(var i=1; i<=4; i++){
    for(var j=1; j<=4; j++){
        if (i === 1) {M[i][j] = Math.round(M[i][j]) / Ta;}
        if (i === 2) {M[i][j] = Math.round(M[i][j]) / Tt;}
        if (i === 3) {M[i][j] = Math.round(M[i][j]) / Tg;}
        if (i === 4) {M[i][j] = Math.round(M[i][j]) / Tc;}
    }
}
}
</script>
```

```
JavaScript Output:
Base(1) = [0.16666666666 | 0.50000000000000 | 0.16666666666667 |
          0.16666666666667]
Base(2) = [0.27380952380 | 0.35317460317460 | 0.17063492063492 |
          0.20238095238095]
Base(3) = [0.22697782816 | 0.35169438145629 | 0.21003401360544 |
          0.21129377676997]
BasesPredicted[TTT]
```

Supporting algorithm (JS alternative) 10.

```
<script>
var M = [];
var v = [];

for(var i=1; i<=4; i++){M[i]=[];}
for(var i=0; i<=3; i++){v[i]=[];}

ExtractProb("TACTTCGATTTAAGCGCGGCGGCCTATATTA");

var chain = 5;
```

```javascript
v[0][0] = 1;
v[1][0] = 0;
v[2][0] = 0;
v[3][0] = 0;

v[0][1] = 0;
v[1][1] = 0;
v[2][1] = 0;
v[3][1] = 0;

for(var k=1; k<=chain; k++){
    for(var i=0; i<=3; i++){
        for(var j=0; j<=3; j++){
            v[i][1] = v[i][1] + (v[j][0] * M[j + 1][i + 1]);
        }
    }

for(var i=0; i<=3; i++){
    v[i][0] = v[i][1];
    v[i][1] = 0;
}

    var A = v[0][0];
    var T = v[1][0];
    var G = v[2][0];
    var C = v[3][0];

document.write("V("+k+")=["+A+"|"+T+"|"+G+"|"+C+"]</br>");
}

function ExtractProb(s) {
var Ea = "A";
var Et = "T";
var Eg = "G";
var Ec = "C";

for(var i=1; i<=4; i++){
    M[i]=[];
    for(var j=1; j<=4; j++){
        M[i][j]=0;
    }
}
```

```
var Ta = 0;
var Tt = 0;
var Tg = 0;
var Tc = 0;
var r;
var c;

for(var i=1; i<s.length-1; i++){

        var DI1 = s.substr(i, 1);
        var DI2 = s.substr(i+1, 1);

        if (DI1 === Ea) {r = 1;}
        if (DI1 === Et) {r = 2;}
        if (DI1 === Eg) {r = 3;}
        if (DI1 === Ec) {r = 4;}

        if (DI2 === Ea) {c = 1;}
        if (DI2 === Et) {c = 2;}
        if (DI2 === Eg) {c = 3;}
        if (DI2 === Ec) {c = 4;}

        M[r][c] = M[r][c] + 1;

        if (DI1 === Ea) {Ta = Ta + 1;}
        if (DI1 === Et) {Tt = Tt + 1;}
        if (DI1 === Eg) {Tg = Tg + 1;}
        if (DI1 === Ec) {Tc = Tc + 1;}
}

for(var i=1; i<=4; i++){
    for(var j=1; j<=4; j++){
        if (i === 1) {M[i][j] = Math.round(M[i][j]) / Ta;}
        if (i === 2) {M[i][j] = Math.round(M[i][j]) / Tt;}
        if (i === 3) {M[i][j] = Math.round(M[i][j]) / Tg;}
        if (i === 4) {M[i][j] = Math.round(M[i][j]) / Tc;}
    }
}
}
</script>
```

```
JavaScript Output:
V(1)=[0.16666666666666|0.50000000000000|0.16666666666666|
      0.16666666666666]
V(2)=[0.27380952380952|0.35317460317460|0.17063492063492|
      0.20238095238095]
V(3)=[0.22697782816830|0.35169438145628|0.21003401360544|
      0.21129377676996]
V(4)=[0.22414311109511|0.33016717857042|0.21857865721244|
      0.22711105312201]
V(5)=[0.21532367589496|0.32370139612165|0.22958597474151|
      0.2313889532418]
```

Supporting algorithm (JS alternative) 11.

```
<script>
var Inp = [];
var Obs = "";
var Reg = "";
var l;

var R =
"159,82,187,194,179,115,197,102,105,104,95,126,74,143,143,
127,98,70,92,170,168,182,149,85,137,100,170,180,61,177,86,
195,198,182,150,197,103,103,186,100,96,196";

Inp = R.split(",");
var Lu = 200;
var Ld = 60;
var n = 4;

Pr = (Lu - Ld) / n;

for(var i=0; i<Inp.length; i++){

    s = (Inp[i] - Ld) / Pr;
    s = Math.floor(s);

    if(s === 0){l = "A";}
    if(s === 1){l = "B";}
    if(s === 2){l = "C";}
    if(s === 3){l = "D";}
```

```
        Obs = Obs + 1;
        Reg = Reg + s + ",";
}


document.write("Reg=" + Reg + "</br>" + "Obs=" + Obs +
"</br>");
</script>
```

JavaScript Output:

```
Reg=2,0,3,3,3,1,3,1,1,1,1,1,0,2,2,1,1,0,0,3,3,3,2,0,2,1,3,3,0,3,
0,3,3,3,2,3,1,1,3,1,1,3,
Obs=CADDDBDBBBBBACCBBAADDDCACBDDADADDDCDBBDBBD
```

Supporting algorithm (JS alternative) 12.

```
<script>
var Jar = [];
var v = [];

for(var i=1; i<=3; i++){Jar[i]=[];}
for(var i=0; i<=3; i++){v[i]=[];}

Jar[1][1] = 0.33;
Jar[1][2] = 0.33;
Jar[1][3] = 0.33;

Jar[2][1] = 0.5;
Jar[2][2] = 0.5;
Jar[2][3] = 0;

Jar[3][1] = 0;
Jar[3][2] = 0;
Jar[3][3] = 1;


var chain = 5;

v[0][0] = 1;
v[1][0] = 0;
v[2][0] = 0;
```

```
v[0][1] = 0;
v[1][1] = 0;
v[2][1] = 0;

for(var k=1; k<=chain; k++){
    for(var i=0; i<=2; i++){
        for(var j=0; j<=2; j++){
            v[i][1] = v[i][1] + (v[j][0] * Jar[j + 1][i + 1]);
        }
    }

for(var i=0; i<=2; i++){
    v[i][0] = v[i][1];
    v[i][1] = 0;
}

    var A = v[0][0];
    var B = v[1][0];
    var C = v[2][0];
document.write("Step("+k+")=["+A+"|"+B+"|"+C+"]</br>");
}
</script>
```

```
JavaScript Output:
Step(1)=[0.33|0.33|0.33]
Step(2)=[0.2739|0.2739|0.4389]
Step(3)=[0.227337|0.227337|0.529287]
Step(4)=[0.18868971|0.18868971|0.60430821]
Step(5)=[0.1566124593|0.1566124593|0.6665758143]
```

Supporting algorithm (JS alternative) 13.

```
<script>
var Jar = [];
var v = [];

for(var i=1; i<=4; i++){Jar[i]=[];}
for(var i=0; i<=3; i++){v[i]=[];}

Jar[1][1] = 1;
Jar[1][2] = 0;
```

```
Jar[1][3] = 0;
Jar[1][4] = 0;

Jar[2][1] = 0.5;
Jar[2][2] = 0;
Jar[2][3] = 0.5;
Jar[2][4] = 0;

Jar[3][1] = 0;
Jar[3][2] = 0.5;
Jar[3][3] = 0;
Jar[3][4] = 0.5;

Jar[4][1] = 0;
Jar[4][2] = 0;
Jar[4][3] = 1;
Jar[4][4] = 0;

var chain = 5;

v[0][0] = 0;
v[1][0] = 0;
v[2][0] = 0;
v[3][0] = 1;

v[0][1] = 0;
v[1][1] = 0;
v[2][1] = 0;
v[3][1] = 0;

for(var k=1; k<=chain; k++){
    for(var i=0; i<=3; i++){
        for(var j=0; j<=3; j++){
            v[i][1] = v[i][1] + (v[j][0] * Jar[j + 1][i + 1]);
        }
    }
}

for(var i=0; i<=3; i++){
    v[i][0] = v[i][1];
    v[i][1] = 0;
}

    var A = v[0][0];
    var B = v[1][0];
```

```
    var C = v[2][0];
    var D = v[3][0];

document.write("Step("+k+")=["+A+"|"+B+"|"+C+"|"+D+"]</br>");
}
</script>
```

```
JavaScript Output:
Step(1)=[0|0|1|0]
Step(2)=[0|0.5|0|0.5]
Step(3)=[0.25|0|0.75|0]
Step(4)=[0.25|0.375|0|0.375]
Step(5)=[0.4375|0|0.5625|0]
```

Supporting algorithm (JS alternative) 14.

```
<script>
var P = [];
var Jar = [];

for(var i=0; i<=3; i++){P[i]=[];}
for(var i=1; i<=3; i++){Jar[i]=[];}

P[0][0] = "A";
P[0][1] = "B";
P[0][2] = "C";

P[1][0] = 0.33;
P[1][1] = 0.33;
P[1][2] = 0.33;

P[2][0] = 0;
P[2][1] = 0.5;
P[2][2] = 0.5;

P[3][0] = 1;
P[3][1] = 0;
P[3][2] = 0;

for(var j=1; j<=3; j++){
    Jar[j] = Fill_Jar(j);
```

```
        if (j>0) {document.write("Jar("+j+")=("+Jar[j]+")</br>");}
    }

    var draws = 20;
    var q = "";
    var z = "";
    var a;

    a = Draw(1);

    for (var i=1; i<=draws; i++) {
        for (var j=0; j<=2; j++) {
            if (a === P[0][j]) {
                a = Draw(j + 1);
                q = q + P[0][j];
                z = z + ", Jar " + P[0][j] + "[" + a + "]";
                j=2;
            }
        }
    }

    document.write("Q = " + q + "</br>");
    document.write("Z = " + z);

    function Draw(S) {
        randomly_choose = Math.floor((Math.random() * Jar[S].
    length));
        ball = Jar[S].substr(randomly_choose, 1);
        return ball;
    }

    function Fill_Jar(S) {
    var Ltot = 27;
    var a = 1;
    var b = "";
        for (var i=0; i<=2; i++) {
        a = Math.round(Ltot * P[S][i]);
            for (var j=1; j<=a; j++) {
                b = b + P[0][i];
            }
        }
    return b;
    }
    </script>
```

```
JavaScript Output:
Jar(1)=(AAAAAAAAABBBBBBBBBBCCCCCCCCC)
Jar(2)=(BBBBBBBBBBBBBBBCCCCCCCCCCCCCCC)
Jar(3)=(AAAAAAAAAAAAAAAAAAAAAAAAAAAAAA)


Q = CABCABBBBCABCABBBBCA


Z = , Jar C[A], Jar A[B], Jar B[C], Jar C[A], Jar A[B],
Jar B[B], Jar B[B], Jar B[B], Jar B[C], Jar C[A], Jar A[B],
Jar B[C], Jar C[A], Jar A[B], Jar B[B], Jar B[B], Jar B[B],
Jar B[C], Jar C[A], Jar A[C]
```

Supporting algorithm (JS alternative) 15.

```
<script>
var P = [];
var Jar = [];

for(var i=0; i<=4; i++){P[i]=[];}
for(var i=1; i<=4; i++){Jar[i]=[];}

P[0][0] = "A";
P[0][1] = "B";
P[0][2] = "C";
P[0][3] = "D";

P[1][0] = 0;
P[1][1] = 1;
P[1][2] = 0;
P[1][3] = 0;

P[2][0] = 0.33;
P[2][1] = 0;
P[2][2] = 0.33;
P[2][3] = 0.33;

P[3][0] = 0;
P[3][1] = 1;
P[3][2] = 0;
P[3][3] = 0;
```

```
P[4][0] = 0;
P[4][1] = 0;
P[4][2] = 1;
P[4][3] = 0;

for(var j=1; j<=4; j++){
    Jar[j] = Fill_Jar(j);
}

var draws = 100;
var q = "";
var z = "";
var a;

a = Draw(1);

for (var i=1; i<=draws; i++){
    for (var j=0; j<=3; j++){
        if (a === P[0][j]){
            a = Draw(j + 1);
            q = q + P[0][j];
            j=3;
        }
    }
}

document.write("Q = " + q + "</br>");

function Draw(S) {
    randomly_choose = Math.floor((Math.random() * Jar[S].
length));
    ball = Jar[S].substr(randomly_choose, 1);
    return ball;
}

function Fill_Jar(S){
var Ltot = 100;
var a = 1;
var b = "";
    for (var i=0; i<=2; i++){
    a = Math.round(Ltot * P[S][i]);
```

```
        for (var j=1; j<=a; j++){
            b = b + P[0][i];
        }
    }
return b;
}
</script>
```

JavaScript Output:

Q =BABABABABABCBCBABABCBABCBABABCBABCBCBCBCBCBABCBCBCBCBCBABABA
BCBCBABABABCBCBCBCBCBABABCBABABABCBABA

Supporting algorithm (JS alternative) 16.

```
<script>
var P = [];

for(var i=1; i<=4; i++){
    P[i]=[];
    for(var j=1; j<=4; j++){
        P[i][j]=0;
    }
}

var z = "";

ExtractProb("BABDCBDCBABDCBCBDCBABABABDCBDCBDCBCBABABDCBD
CBDCBABCBABCBDCBDCBDCBDCBABCBCBABABDCBDCBCBDCBCBDCBDCBAB");

z = "<table border='1'>";
for(var i=1; i<=4; i++){
    z += "<tr>";
    for(var j=1; j<=4; j++){
        z += "<td>" + P[i][j].toFixed(2) + "</td>" ;
    }
z += "</tr>";
}
z += "</table>";

document.write(z);
```

```
function ExtractProb(s) {
var Ea = "A";
var Eb = "B";
var Ec = "C";
var Ed = "D";

var Ta = 0;
var Tb = 0;
var Tc = 0;
var Td = 0;
var r;
var c;

for(var i=1; i<s.length-1; i++){

        var DI1 = s.substr(i, 1);
        var DI2 = s.substr(i+1, 1);

        if (DI1 === Ea) {r = 1;}
        if (DI1 === Eb) {r = 2;}
        if (DI1 === Ec) {r = 3;}
        if (DI1 === Ed) {r = 4;}

        if (DI2 === Ea) {c = 1;}
        if (DI2 === Eb) {c = 2;}
        if (DI2 === Ec) {c = 3;}
        if (DI2 === Ed) {c = 4;}

        P[r][c] = P[r][c] + 1;

        if (DI1 === Ea) {Ta = Ta + 1;}
        if (DI1 === Eb) {Tb = Tb + 1;}
        if (DI1 === Ec) {Tc = Tc + 1;}
        if (DI1 === Ed) {Td = Td + 1;}
}

for(var i=1; i<=4; i++){
    for(var j=1; j<=4; j++){
        if (i === 1) {P[i][j] = Math.round(P[i][j]) / Ta;}
        if (i === 2) {P[i][j] = Math.round(P[i][j]) / Tb;}
        if (i === 3) {P[i][j] = Math.round(P[i][j]) / Tc;}
```

```
          if (i === 4) {P[i][j] = Math.round(P[i][j]) / Td;}
     }
  }
}
</script>
```

JavaScript Output:			
0	1	0	0
0.31	0	0.21	0.49
0	1	0	0
0	0	1	0

Supporting algorithm (JS alternative) 17.

```
<script>
var Jar = [];
var P = [];

for(var i=0; i<=4; i++){
    P[i]=[];
    for(var j=0; j<=3; j++){
        P[i][j]=0;
    }
}

var f = [];
var pro = "";

f[0] = 0;
f[1] = 0;
f[2] = 0;
f[3] = 0;

P[0][0] = "A";
P[0][1] = "B";
P[0][2] = "C";
P[0][3] = "D";

P[1][0] = 0;
P[1][1] = 1;
P[1][2] = 0;
P[1][3] = 0;
```

```
P[2][0] = 0.33;
P[2][1] = 0;
P[2][2] = 0.33;
P[2][3] = 0.33;

P[3][0] = 0;
P[3][1] = 1;
P[3][2] = 0;
P[3][3] = 0;

P[4][0] = 0;
P[4][1] = 0;
P[4][2] = 1;
P[4][3] = 0;

for(var j=1; j<=4; j++){
    Jar[j] = Fill_Jar(j);
}

var draws = 10000;
var z = "";
var a;

a = Draw(2);

for (var i=1; i<=draws; i++){
    for (var j=0; j<=3; j++){
        if (a === P[0][j]){
            a = Draw(j + 1);
            z += a;
            j=3;
        }
    }
}

for(var i=1; i<z.length; i++){
    g = z.substr(i, 1);
    if(g === "A"){f[0] = f[0] + 1;}
    if(g === "B"){f[1] = f[1] + 1;}
    if(g === "C"){f[2] = f[2] + 1;}
    if(g === "D"){f[3] = f[3] + 1;}
}
```

```
for(var i=0; i<=3; i++){
pro += P[0][i] + "=" + Math.round((100 / z.length) * f[i]) +
"% ";
}

document.write(pro);

function Draw(S) {
    randomly_choose = Math.floor((Math.random() * Jar[S].
length));
    ball = Jar[S].substr(randomly_choose, 1);
    return ball;
}

function Fill_Jar(S){
var Ltot = 100;
var a = 1;
var b = "";
    for (var i=0; i<=3; i++){
    a = Math.round(Ltot * P[S][i]);
        for (var j=1; j<=a; j++){
            b = b + P[0][i];
        }
    }
return b;
}
</script>
```

JavaScript Output:

A=14%   B=43%   C=29%   D=14%

C

Syntax Equivalence between Languages

Markov Chains: From Theory to Implementation and Experimentation, First Edition. Paul A. Gagniuc.
© 2017 John Wiley & Sons, Inc. Published 2017 by John Wiley & Sons, Inc.
Companion website: www.wiley.com/go/gagniuc/markovchains

JavaScript	Visual Basic	PHP
`function Fill_Jar(S){`	`Function Fill_Jar(ByVal S As Variant)` `As Variant`	`function Fill_Jar($S){`
`}`	`End Function`	`}`
`var P = [];`	`Dim P(0 To 4, 0 To 3) As Variant`	`$P = array();`
`P[S][i]`	`P(S, i)`	`$P[$S][$i]`
`Math.round(Ltot * P[S][i]);`	`Int(Ltot * P(S, i))`	`round($Ltot * $P[$S][$i]);`
`.substr(randomly_choose, 1);`	`Mid(Jar(S), randomly_choose + 1, 1)`	`substr($Jar[$S], $randomly_choose, 1);`
`for (var i=0; i<=3; i++){`	`For i = 0 To 3`	`for ($i=0; $i<=3; $i++){`
`}`	`Next i`	`}`
	`Randomize`	`srand(mktime());`
`return b;`	`Draw = ball`	`return $b;`
`if (a === P[0][j]){`	`If a = P(0, j) Then`	`if ($a == $P[0][$j]){`
`} else {`	`Else`	`} else {`
`}`	`End If`	`}`
`alert("Jar");`	`MsgBox "Jar"`	`echo "Jar";`
`Jar[S].strlen`	`Len(Jar(S))`	`strlen($Jar[$S])`
`Math.floor((Math.random() *` `Jar[S].length));`	`Int(Rnd * Len(Jar(S)))`	`rand(1,strlen($Jar[$S])-1);`
`b = b + P[0][i];`	`b = b & P(0, i)`	`$b = $b . $P[0][$i]; or $b .= $P[0][$i];`
`s = Math.floor(s);`	`s = Split(s, ",")(0)`	`$s = floor($s);`
`Inp = R.split(",");`	`Inp = Split(R, ",")`	`$Inp = explode(",", $R);`
`Inp.length`	`UBound(Inp)`	`count($Inp)`

Glossary

Probability Indicates the behavior of a stochastic process.

Probability vs. Statistics Probability deals with random processes behind outcomes and statistics deals with inferences from data.

Stochastic Process A random process that evolves over time. A stochastic process and a random process have often the same meaning.

Random Process The evolution of a system represented by a random Variable.

Random Variable A variable whose value is subject to variations due to chance.

Discrete Variable A variable that has a finite number of possible values with no inherent order. The hand of a clock that moves from second to second or from minute to minute is a good example of a discrete variable (in this case, there is an order, however). Discrete variables are also known as categorical variables or qualitative variables.

Continuous Variable Opposing to discrete variables are continuous variables that can take an infinite number of possible values in a certain range. For instance, time and distance are continuous variables. However, a measured variable is not truly a continuous variable. For instance, a measuring device has a limited number of digits and therefore, does not allow for an infinite number of possible values in a certain range.

Discrete Random Variable A discrete random variable is a random variable that can take only a finite set of outcomes.

Outcome Represents a possible result of an experiment. In the case of a dice, the outcome may be any number between 1 and 6.

Experiment An experiment is represented by a method that can be infinitely repeated and has a defined set of possible outcomes. For instance, an experiment can be the throwing of a dice 10,000 times while observing each result. Each throw would be considered a trial within the experiment composed of 10,000 throws.

Sample Space Represents the set of all possible outcomes of an experiment. In the case of a dice, the sample space consists of 1, 2, 3, 4, 5, 6.

Markov Chains: From Theory to Implementation and Experimentation, First Edition. Paul A. Gagniuc.
© 2017 John Wiley & Sons, Inc. Published 2017 by John Wiley & Sons, Inc.
Companion website: www.wiley.com/go/gagniuc/markovchains

Event An event is a set of outcomes of an experiment. A single outcome may be an element of many different events.

State Diagram State diagrams are visual representations that are used to give an abstract description of the behavior of a system.

Machine An abstract concept which refers to a system such as a Markov chain. For instance, the process of rolling a dice is referred to as a "machine".

Discrete-Time Markov Chain Discrete-time Markov chain (DTMC) represents a random process that undergoes transitions from one state to another state on a state space.

State Space A state space is the representation of all possible states of a system.

Discrete Steps The value of a discrete random variable changes only at certain intervals (separate points in time) called discrete steps. For instance, a discrete step is made when a dice stops after being thrown. If thrown again and the dice stops, then the second discrete step is made.

Probability Distribution The probability distribution of a discrete random variable is composed of a list of probabilities, each associated with an outcome of a random experiment.

Probability Vector A probability vector is a list of probability values that represent the possible outcomes of a discrete random variable. A probability vector or a stochastic vector is a vector with non-negative entries that add up to one.

Vector Components A component of a vector denotes a probability value that represents one of the possible outcomes of a discrete random variable.

Average Time It shows how often a state is visited on average if the number of discrete steps tends to infinity.

$P(A)$ **or** $P[A]$ It represents a probability function. The meaning of $P(A)$ or $P[A]$ is "the probability of event A".

$P(A|B)$ The classic notation of a conditional probability function. The meaning of $P(A|B)$ is "probability of event A given event B occurred". In other words, it signifies the probability of transition from state "A" to state "B".

$P[A|B]$ Denotes the transition probability from a state "A" to another state "B". A transition probability can be also represented by a conditional probability function, where $P(B|A)$ is equivalent to $P[A|B]$.

References

1 Bellhouse, D. (2005). Decoding Cardano's Liber de Ludo Aleae. *Hist. Math.*, 32:180–202.

2 Bellhouse, D. R. (2000). De Vetula: a medieval manuscript containing probability calculations. *Int. Stat. Rev.*, 68:123–136.

3 Franklin, J. (2001). *The Science of Conjecture: Evidence and Probability before Pascal*. Baltimore, MD: Johns Hopkins University Press. ISBN 0-8018-6569-7.

4 Hacking, I. (2006). *The Emergence of Probability*, 2nd ed. New York: Cambridge University Press. ISBN 978-0-521-86655-2.

5 Hald, A. (2003). *A History of Probability and Statistics and Their Applications before 1750*. Hoboken, NJ: John Wiley & Sons. ISBN 0-471-47129-1.

6 Bernoulli, J. (1899). *Wahrscheinlichkeitsrechnung* (Ars Conjectandi, 1713). Ostwalds Klassiker der Exakten Wissenschaften. Leipzig, Germany: Wilhelm Engelmann.

7 Seneta, E. (2013). A tricentenary history of the law of large numbers. *Bernoulli*, 19(4):1088–1121.

8 Laplace, P. (1812). *Théorie Analytique des Probabilités*. Paris: Courcier.

9 Molina, E. C. (1930). The theory of probability: some comments on Laplace's théorie analytique. *Bull. Amer. Math. Soc.*, 36:369–392.

10 Basharin, G. P., Langville, A. N., and Naumov, V. A. (2004). The life and work of A. A. Markov. *Linear Algebra Appl.*, 386:3–26.

11 Марков, А. А. (1906). Распространение закона больших чисел на величины, зависящие друг от друга. Известия Физико-математического общества при Казанском университете, 2-я серия, том 15, ст. 135–156 [Translation: Spreading the law of large numbers by quantities that depend on each other].

12 Hayes, B. (2013). First links in the Markov chain. *Am. Sci.*, 101(2): 92–97.

13 Cohen, J. E., Kemperman, J. H. B., and Zbaganu, G. (1998). *Comparisons of Stochastic Matrices, with Applications in Information Theory, Statistics, Economics and Population Sciences*. Boston, Basel, and Berlin: Birkhäuser.

14 de Matthew, H., and Petoukhov, S. (2011). *Mathematics of Bioinformatics: Theory, Practice and Applications*. New York: Wiley-Interscience.

Markov Chains: From Theory to Implementation and Experimentation, First Edition. Paul A. Gagniuc.
© 2017 John Wiley & Sons, Inc. Published 2017 by John Wiley & Sons, Inc.
Companion website: www.wiley.com/go/gagniuc/markovchains

15 Gillespie, D. T. (1976). A general method for numerically simulating the stochastic time evolution of coupled chemical reactions. *J. Comput. Phys.*, 22:403–434.

16 Ribeiro, A. S., Zhu, R., and Kauffman, S. A. (2006). A general modeling strategy for gene regulatory networks with stochastic dynamics. *J. Comp. Biol.*, 13(9):1630–1639.

17 Bapat, R. B., and Raghavan, T. E. S. (1997). *Nonnegative Matrices and Applications*. Cambridge: Cambridge University Press.

18 Wilks, D. S., and Wilby, R. L. (1999). The weather generation game: a review of stochastic weather models. *Prog. Phys. Geogr.*, 23(3):329–357.

19 Schoof, J. T., and Pryor, S. C. (2008). On the proper order of Markov chain model for daily precipitation occurrence in the contiguous United States. *J. Appl. Meteor. Climatol.*, 47:2477–2486.

20 Gabriel, K. R., and Neumann, J. (1962). A Markov chain model for daily rainfall occurrence at Tel Aviv. *Q. J. R. Meteorol. Soc.*, 88:90–95.

21 Chin, E. H. (1977). Modeling daily precipitation occurrence process with Markov chain. *Water Resour. Res.*, 13:949–956.

22 Katz, R. W. (1977). Precipitation as a chain-dependent process. *J. Appl. Meteor.*, 16:671–676.

23 Carey, D. I., and Haan, C. T. (1978). Markov processes for simulating daily point rainfall. *J. Irrig. Drain. Div. ASCE*, 104(IR1):111–125.

24 Haan, C. T., Allen, D. M., and Street, J. O. (1976). A Markov chain model of daily rainfall. *Water Resour. Res.*, 12(3):443–449.

25 Sonnadara, D. U. J., and Jayewardene, D. R. (2015). A Markov chain probability model to describe wet and dry patterns of weather at Colombo. *Theor. Appl. Climatol.*, 119(1–2):333–340.

26 Kitchin, R. (2014). *The Data Revolution: Big Data, Open Data, Data Infrastructures and Their Consequences*. London: Sage.

27 Kitchin, R. (2014). Big Data, new epistemologies and paradigm shifts. *Big Data Soc.*, 1(1):1–12.

28 Miller, H. J. (2010). The data avalanche is here. Shouldn't we be digging? *J. Regional Sci.*, 50(1):181–201.

29 Floridi, L. (2012). Big data and their epistemological challenge. *Philos. Technol.*, 25(4):435–437.

30 Han, J., Kamber, M., and Pei, J. (2011). *Data Mining: Concepts and Techniques*, 3rd ed. Waltham: Morgan Kaufmann.

31 Ramsay, S. (2010). *Reading Machines: Towards an Algorithmic Criticism*. Champaign: University of Illinois Press.

32 Leonelli, S. (2012). Introduction: making sense of data-driven research in the biological and biomedical sciences. *Stud. Hist. Philos. Biol. Biomed. Sci.*, 43(1):1–3.

33 Strasser, B. J. (2012). Data-driven sciences: from wonder cabinets to electronic databases. *Stud. Hist. Philos. Biol. Biomed. Sci.*, 43:85–87.

34 Stephens, Z. D., Lee, S. Y., Faghri, F., Campbell, R. H., Zhai, C., Efron, M. J., Iyer, R., Schatz, M. C., Sinha, S., and Robinson, G. E. (2015). Big Data: astronomical or genomical? *PLoS Biol.*, 13:e1002195.

35 Gagniuc, P. A., and Ionescu-Tirgoviste, C. (2013). Gene promoters show chromosome specificity and reveal chromosome territories in humans. *BMC Genom.*, 14:278.

36 Gagniuc, P. A., and Ionescu-Tirgoviste, C. (2012). Eukaryotic genomes may exhibit up to 10 generic classes of gene promoters. *BMC Genom.*, 13:512.

37 Churchill, G. A. (1989). Stochastic models for heterogeneous DNA sequences. *Bull. Math. Biol.*, 51(1):79–94.

38 Durbin, R., Eddy, S. R., Krogh, A., and Mitchison, G. (1998). *Biological Sequence Analysis*. Cambridge University Press.

39 Ionescu-Tirgoviste, C., Gagniuc, P. A., Gubceac, E., Mardare, L., Popescu, I., Dima, S., and Militaru, M. (2015). A 3D map of the islet routes throughout the healthy human pancreas. *Sci. Rep.* 5:14634.

40 Pais, I., Hallschmid, M., Jauch-Chara, K., Schmid, S. M., Oltmanns, K. M., Peters, A., Born, J., and Schultes, B. (2007). Mood and cognitive functions during acute euglycaemia and mild hyperglycaemia in type 2 diabetic patients. *Exp. Clin. Endocrinol. Diabetes*, 115(1):42–46.

41 Sommerfield, A. J., Deary, I. J., and Frier, B. M. (2004). Acute hyperglycemia alters mood state and impairs cognitive performance in people with type 2 diabetes. *Diabetes Care*, 27(10):2335–2340.

42 Capes, S. E., Hunt, D., Malmberg, K., Pathak, P., and Gerstein, H. C. (2001). Stress hyperglycemia and prognosis of stroke in nondiabetic and diabetic patients: a systematic overview. *Stroke*. 32(10):2426–2432.

43 Cryer, P. E., Axelrod, L., Grossman, A. B., Heller, S. R., Montori, V. M., Seaquist, E. R., and Service, F. J. (2009). Evaluation and management of adult hypoglycemic disorders: an Endocrine Society Clinical Practice Guideline. *J. Clin. Endocrinol. Metab.*, 94(3):709–728.

44 Reitsma, J. B., Rutjes, A. W., Whiting, P., Vlassov, V. V., Leeflang, M. M., and Deeks, J. J. (2008, December). Systematic reviews of diagnostic test accuracy. *Ann. Intern. Med.*, 149(12):889–897.

45 Nelis, L. C., and Wootton, J. T. (2010). Treatment-based Markov chain models clarify mechanisms of invasion in an invaded grassland community. *Proc. R. Soc. Lond. B Biol. Sci.*, 277:539–547.

46 Lusseau, D. (2003). Effects of tour boats on the behavior of bottlenose dolphins: using Markov chains to model anthropogenic impacts. *Conserv. Biol.*, 17(6):1785–1793.

47 von Hilgers, P., and Langville, A. N. (2006). The five greatest applications of Markov chains. In: Proceedings of the Markov Anniversary Meeting, Charleston SC, pp. 155–168.

Index

a

absorbing Markov chains
  convergence path   97
  simplistic timeline simulation   93
  state diagrams   93–97, 100
    of cases   127–128
absorbing state, example   94
actual average time   159
Adenine (A)   71, 75
algorithm implementations   37
analysis of frequency   154–155
  average time tester   157
  frequency calculation   155–156
  frequency vs. prediction   157–159
  step-by-step analysis   159
average time spent
  column of matrix   117–119,
    121–122
  cyclical behavior   117
  exemplification of   101
    proportions of   102
  global values, comparison   105
  prediction method   104
  probability matrices   99–100
    occurrence in sequence   119
  recurrent states   126
  sequence of observations   125
  state diagram   99, 103–106, 120
    cases   126
    in particular   100–101

  step-by-step reasoning   120
  transient system   125–126
  transition matrix   122
    cases   123–125
    multiplication   120
average time tester   157

b

Bernoulli, Jakob   1–2
  joint probability of four events   5
  to Markov process   3, 5–8
Bernoulli model   2, 4
blood glucose   85
  down/upper limit   85–86
  four-state Markov chain case   77,
    82
  from measurements to events
    88
blood sugar   84

c

Cardano, Gerolamo   2
computational tools   61
computation methods   61
computer code   68
  implementation   95–96
  three-state Markov chain issue
    68
cyclic permutations   74
Cytosine (C)   71, 75

d

de Fermat, Pierre 1
dependent events 2
dependent variables 2–5
De Ratiociniis in Ludo Aleae, 1
diabetes 84
digital representations 74
discrete-time Markov chain (DTMC)
    71, 84
DNA sequence
  Markov diagram 71
  nucleotides 72
  step-by-step prediction 77–80
double stochastic matrix 58
DTMC, *see* discrete-time Markov
    chain (DTMC)

e

electronic format, representation 18
expected average time 154, 157
ExtractProb function 84, 153

f

four-state diagrams, cases 117–123
frequency analysis 161

g

Gambler's Ruin, cases 128–130
glycemic values 87
  conversion of measurements to
    states 91
  from measurements 85
  Reg/Obs variables 91
    sequence of observations 90
  total number of states 88
Guanine (G) 71
Guanine state 75

h

"heavy rain" 132

i

identity matrix 50
initial probability vector 65
insulin hormone 84

j

jar content 144–145
"Jar" vector 149–150
JavaScript
  supporting algorithms
    (JS alternative) 1 193–194
  supporting algorithm
    (JS alternative) 2 194–195
  supporting algorithm
    (JS alternative) 3 195–197
  supporting algorithm
    (JS alternative) 4 197–198
  supporting algorithm
    (JS alternative) 5 198–199
  supporting algorithm
    (JS alternative) 6 199–200
  supporting algorithm
    (JS alternative) 7 200–201
  supporting algorithm
    (JS alternative) 8 202–203
  supporting algorithm
    (JS alternative) 9 204–206
  supporting algorithm
    (JS alternative) 10 206–209
  supporting algorithm
    (JS alternative) 11 209–210
  supporting algorithm
    (JS alternative) 12 210–211
  supporting algorithm
    (JS alternative) 13 211–213
  supporting algorithm
    (JS alternative) 14 213–215
  supporting algorithm
    (JS alternative) 15 215–217
  supporting algorithm
    (JS alternative) 16 217–219
  supporting algorithm
    (JS alternative) 17 219–221

l

Liber de Ludo Aleae 1

m

Markov, Andrei 2
Markov chains 19, 25, 80, 93, 107

absorbing, *see* absorbing Markov chains
analysis of different settings   57
case   77
double stochastic matrix   56, 58
four-state
  computer code   82–83
  predictions   71–80
  vectors   81
graphical representation of   42
implementations of   61
jars and balls   133
linguistic analysis   5
method   84
modeling on measurements   84–91
n-state
  predictions   80–84
  predictions based on sequences   83
probabilities of two jars   8
probability vectors   81
right stochastic matrix   59
simulation of   131–163
  algorithm for   132
steady-state vector   56
stochastic matrix   58
technological applications   99
three-state   62, 74, 132
  convergence path   105
  predictions   61–71
transition matrix   133
two-state   7, 9, 33, 61, 74
  long-run distribution   55–59
  cases   56–59
  n-state   61
  parameters for   141–144
  predictions, performing   37–46
  simulation of   14–24
  new jars, representation of   17–19
  original machine, black jar of   17
  original machine, white jar of   17

system   19–24
  steady state of   46–55
  transition probabilities of   33
Markov chain simulator
  based on probability values   23
  expansion for   149
  framework for   148
  software implementation   59
  two-state   21
Markov diagram   13–14, 37
  to Markov matrix   10
  observed sequence   73
  for sequence   38
    observations   15–16
  to jars and balls   133–134
  weather-related sequence of observations   29
Markovian manner   84
Markov machine
  four-state
    DNA/RNA, sequences   70
    protein sequences   70
  three-state
    step-by-step prediction, using observations sequence   70
  two-state
    long-run distribution of   56
    step-by-step prediction   52–55
Markov matrix   9, 13–14
Markov process   62
matrix, *see* specific matrices
matrix framework   149–150
matrix multiplication, probability vectors   81

n
"no rain"   132
nucleotide prediction   76–77
nucleotides   71

o
output verification   150–154
  transition probabilities calculation   150–151
  transition probability tester   153

p

Pascal, Blaise   1
PHP
   supporting algorithm
      (PHP alternative) 1   165–166
   supporting algorithm
      (PHP alternative) 2   166–167
   supporting algorithm
      (PHP alternative) 3   167–169
   supporting algorithm
      (PHP alternative) 4   169–170
   supporting algorithm
      (PHP alternative) 5   170–171
   supporting algorithm
      (PHP alternative) 6   171
   supporting algorithm
      (PHP alternative) 7   171–173
   supporting algorithm
      (PHP alternative) 8   173–175
   supporting algorithm
      (PHP alternative) 9   175–177
   supporting algorithm
      (PHP alternative) 10   178–180
   supporting algorithm
      (PHP alternative) 11   180–181
   supporting algorithm
      (PHP alternative) 12   181–182
   supporting algorithm
      (PHP alternative) 13   182–184
   supporting algorithm
      (PHP alternative) 14   184–185
   supporting algorithm
      (PHP alternative) 15   186–187
   supporting algorithm
      (PHP alternative) 16   187–189
   supporting algorithm
      (PHP alternative) 17   189–191
prediction rules   39
probability, *see* specific probabilities
probability formula, mathematical
   theory of   1
probability functions, conditional   26,
   33–34

probability matrices   9
   probability vectors   99
probability value   43, 146
   Markov chain framework, simulation
      149
probability vectors   33–34, 39–41, 50,
   68, 76, 81–82
   component   74
   hypothetical sequence   44
   second probability value   40
   vector components, step-by-step
      calculation of   66–68
problem solving   116
protein sequences   80

r

"rain"   132
random processes behavior   159–163
   increasing trends   161
   observations vs. expectations   160
   overlapping distribution   162
rounding down, value   89

s

simulation behavior   131–132
   imbalance of chances   135–137
   imposed ratios   138
   Markov diagrams to jars and balls
      133–134
   representation of the jars   134–135
   simulation of the system   139–141
   state diagram   132
simulation of chain configurations
   145–150
   alternative parameters   145–146
   framework expansion   145, 148
state diagrams
   with absorbing states   127–128
   with associated probabilities   102
   of cases   123–127
steady-state vector   37, 46
stochastic matrices   9, 25, 77
   building from events   25–32

building from percentages   32–35
doubly   9
probability theory   9–11
types of   11
stochastic process, transition
        probability   12
stroke   84

t

Théorie Analytique des Probabilités
    2
three-state diagrams, case   115–117
three-state Markov chain simulator
        143–144
Thymine   71, 76
transition matrix   9, 14, 17, 30, 40–41,
        55, 57, 64, 75, 95, 132, 153
    absorbing state   127
    columns of   100
    elements of   75
    hypothetical sequence   45
        of observations   44
    from known events   26
    multiplications of matrix   41, 101
    probability of rainy day   41, 65
    probability of sunny day   43
    probability values   38
    probability vector   46, 57, 59
    sequence of observations   15–16
    series of cases   20–21
    state diagrams   102, 113–116, 118,
        120–123
        with different configurations
            108–110
    steady-state vector   47, 57

2 × 2 transition matrix
    algorithms   53
    probability matrix   28
    steady-state vector   49–52
        computation of   53
    step-by-step prediction   45,
        47–49
    vector multiplication   50
3 × 3 transition matrix   63
    prediction based   96
4 × 4 transition matrix
    prediction framework   112
    two components   58
    unknown values   34–35
    vector components, step-by-step
        calculation of   66
    weather prediction   45–46
    weather-related sequence of
        observations   29
transition probabilities   11–14,
        25–27, 34, 37, 62, 64, 72, 112,
        117
    assumption   93
    matrix   13, 50
    observations   64, 73
    from probability vector   33
    replacing the values   28
    sequence of observations   30–32
    values   27
two square matrices, multiplication of
        103–104
two-state diagrams, cases   113–115

u

unknown probability   33